D0880357

Language and Communication in the Mathematics Classroom

Editors

Heinz Steinbring
University of Dortmund
Dortmund, Germany

Maria G. Bartolini Bussi
University of Modena
Modena, Italy

Anna Sierpinska
Concordia University
Montreal, Quebec

NATIONAL COUNCIL OF TEACHERS OF MATHEMATICS
Reston, Virginia

Copyright © 1998 by
THE NATIONAL COUNCIL OF TEACHERS OF MATHEMATICS, INC.
1906 Association Drive, Reston, VA 20191-1593

Library of Congress Cataloging-in-Publication Data:

Language and communication in the mathematics classroom / editors,
    Heinz Steinbring, Maria G. Bartolini Bussi, Anna Sierpinska.
       p. cm.
    Includes bibliographical references and index.
    ISBN 0-87353-441-7 (pbk.)
      1. Mathematics—Study and teaching. 2. Communication in
education. I. Steinbring, Heinz. II. Bartolini Bussi, Maria G.
(Maria Giuseppina) III. Sierpinska, Anna.
QA11.L3765   1998
510'.71—dc21   98-24453
                                                                       CIP

Printed in the United States of America

Acknowledgments

Many people have contributed to the final version of this book—the authors, of course, but also the participants of the Quebec ICME in 1992 whose feedback was invaluable toward the final output of these chapters.

The preparation of the final version of the manuscript would have been impossible without the financial help of many institutions, mentioned in acknowledgments in specific chapters. In particular, we wish to thank the Canadian and Quebec funding agencies, SSHRC (grant no. 410-93-0007) and FCAR (grant no. 93 ER 1535).

Last but not least, we wish to express our gratitude to Astrid Defence, a doctoral student at Université de Montréal, who not only helped us polish the manuscript from the linguistic and stylistic points of view, but made many pertinent remarks to the editors with regard to the contents of the chapters.

Introduction

THE way we communicate with our students partly determines what we communicate. Let us imagine two classroom situations, both related to teaching students the number fact that 7 plus 8 is 15. In the first situation, the teacher asks a particular student, "Student A, how much is 7 plus 8?" Student A answers, "14." The teacher says, "Wrong!" and turns to another student with the same question. This student answers, "15." The teacher says, "Right!" and goes on to teach students another number fact. In the second situation, the teacher addresses the question, "How much is 7 plus 8?" to the whole class and awaits spontaneous answers. The answers range from 14 to 16. The teacher asks students who have given different answers to explain how they got them. Sometimes, when explaining their strategies, the students whose answer was 14 or 16 recognize by themselves that they made a mistake and speculate on how it could have happened. Sometimes other students point out to them the erroneousness of their strategies and propose changes. The teacher closes the episode by summarizing the various strategies for calculating 7 plus 8 and starts a discussion on how these strategies could be extended to formulate some reliable methods of adding numbers whose sum is greater than 10.

The communication in the first situation followed the pattern of "interrogation": teacher's question→student's answer→teacher's evaluation of the answer. In the second situation, the interaction had the format of a purposeful discussion in which different arguments were proposed, criticized, and compared and a conclusion with regard to a mathematical statement was reached. In the first situation, the criticism was targeted at a particular student, at his or her knowledge and intelligence. In the second situation, the criticism was directed toward various mathematical statements. In both situations, students learned that 7 plus 8 is 15. But much more was learned indirectly. In the first, students learned that knowing mathematics means knowing certain "facts." In the second, students learned that in mathematics, a good

1

argument is worth more than a good memory for facts and that doing mathematics means, among other things, finding regularities, justifying them, and generalizing them.

What do we teach our students if we communicate with them in a particular way? Answering this question is an important issue in mathematics education. Conversely, one can ask, Given a set of instructional goals in teaching mathematics, what kinds of communication can support or hinder the achievement of these goals? These two questions focus on what we might call the "style of communication." But questions arise also with respect to the "means of communication." In mathematics education, these means range from ordinary language to a technical mathematical language, from the language of gestures and icons to the language of symbols, from a literal use of words to a metaphorical use of words, from manipulating material objects to speaking about the possible outcomes of an imagined action on these objects. Every means of communication can support students' understanding of mathematics in some ways and can hinder it in others. To understand exactly how these means of communication function for teachers and students—what makes something an obstacle and what makes it a prop for the learning of mathematics—is also a major task of research in mathematics education.

This book contributes to these issues by building on a series of papers whose first versions were presented in 1992 at the Sixth International Congress of Mathematics Education in Quebec. The analyses of episodes of communication in real mathematics classrooms that can be found in these papers are evidence of both the complexity of these issues and the necessity of developing theoretical and methodological tools for their study, scientific understanding, and explanation. The presentation of these papers is complemented by two new contributions from Anna Sierpinska and Heinz Steinbring.

The book ends with an epilogue that reflects on how the chapters have enriched our understanding of the phenomena of communication in the mathematics classroom and in what way they have informed the theoretical frames of references that underlie our understanding of these phenomena. This concluding chapter also indicates new avenues of study opened up by the contributions to the book.

The book has four parts. The first part, containing the chapters by Pirie and Sierpinska, sets the stage for discussing the topic of language and communication in mathematics education. Pirie's chapter especially underscores the need for theories and methodological tools by highlighting and pointing to the sources of the phenomenal difficulty of

communicating in the mathematical domain. Sierpinska's chapter puts the contributions to the book in the perspective of three general approaches to studying language and communication in the mathematics classroom: constructivism, sociocultural approaches, and interactionism.

Part 2 looks at different approaches to studying communication in mathematics: from the point of view of the sociocultural environment (Bartolini Bussi and Seeger); from the points of view of the "subject matter" to teach, the social interactions, and epistemology (Steinbring); and from the point of view of the use of language (Kanes).

Part 3 offers a window on different styles and patterns of communication in the mathematics classroom. Contrasts between learning environments are discussed in the chapters by Abele, Cestari, and Wood. Communication and learning in small groups are investigated in chapters by Curcio and Artzt, Stacey and Gooding, and Civil. Krummheuer analyzes classroom interactions from the point of view of the theory of formats of argumentation, inspired by Bruner's concept of interaction formats. An alternative pattern of interaction is proposed and studied by Loska.

Part 4 deals with problems of communication that are specific to a variety of mathematical contexts: algebra (Arzarello and MacGregor), functions (Kaldrimidou and Ikonomou), statistics (Clark), logic (Navarra), and metamathematical issues (Fonzi and Smith).

Part 1
Setting the Stage

1

Crossing the Gulf between Thought and Symbol: Language as (Slippery) Stepping-Stones

Susan E. B. Pirie

University of Oxford

Die Mathematiker sind eine Art Franzosen: redet man zu ihnen, so übersetzen sie es in ihre Sprache, und dann ist es alsbald etwas ganz anderes.

—Goethe

THIS chapter does not set out to discuss the genuine research findings or the myths that abound on the subject of communication and mathematical meaning, nor does it aim to give an overview of current work in the area. The intention is to raise some of the problems associated with language that are pertinent and peculiar to the understanding of *mathematics* and to confront some of the received wisdom on the subject. It is a starting point for thinking and for exploring the problematic issues of communication in a mathematics classroom. On the whole, the chapter is concerned only with the communication *of* mathematics and not with the extended field encompassing communication *about* mathematics, although this latter area can be addressed by other authors in this volume.

The starting point of this chapter is that the links between classroom language and pupils' mathematical understandings are tenuous

---

I am indebted to Gontran Ervynck for introducing me to the quotation from Goethe. The English translation follows: Mathematicians are like the French; if you talk to them, they translate it into their own language, and instantly it becomes something quite different.

and tortuous at best and that it is difficult, if not impossible, to establish between them direct, let alone causal, relationships that can be consistently relied on. Language in its broadest sense is the mechanism by which teachers and pupils alike attempt to express their mathematical understandings to each other. It is well accepted that individuals construct understandings that differ not only from one another but that are likely to differ also from the meaning intended by the originator of a particular communication (Pirie and Kieren 1992). The phrase "language in its broadest sense" is used because the communication of mathematics can take place through a variety of forms. This chapter looks at this variety and, through a presentation of classroom incidents, illustrates some of the problems inherent in each form.

MEANS OF MATHEMATICAL COMMUNICATION

The means of mathematical communication can be classified under six headings:

- *Ordinary language.* Here the term *ordinary* denotes the language current in the everyday vocabulary of any particular child, which will, of course, vary for pupils of different ages and stages of understanding.
- *Mathematical verbal language.* *Verbal* here means "using words," either spoken or written.
- *Symbolic language.* This type of communication is made in written, mathematical symbols.
- *Visual representation.* Although not strictly a "language," this is certainly a powerful means of mathematical communication.
- *Unspoken but shared assumptions.* Again, these do not really fall within the definition of "language," but they are a means by which mathematical understanding is communicated and on which new understanding is created. They are ignored at one's peril.
- *Quasi-mathematical language.* This language—usually, but not exclusively, that of the pupils—has, for them, a mathematical significance not always evident to an outsider (who may well, in this context, be the teacher).

Each one of these means of communicating mathematics is legitimate, and each is, and indeed should be, present in any mathematics

classroom. Each of them, in different ways, affects learning and the growth of mathematical understanding of pupils.

FROM THE ORDINARY TO THE MATHEMATICAL LANGUAGE

What implications does this variety of means of communication hold for the understanding of mathematical concepts that pupils build?

First, it could be contended that mathematics has a unique communication problem that arises because the language used when talking about mathematics and that used when writing mathematics (as opposed to writing *about* mathematics) are completely different. A brief but effective illustration can be had by considering this question: What is the difference between 20 and 2? At least three types of legitimate, mathematical, verbal responses are possible:

- "One has two digits, the other has only one" or "One has a zero on the end." Both responses are at a visual level of interpretation of the written mathematical symbolism.

- "They differ by a factor of ten." This is arguably a more mathematical level of interpretation of the question, which leads to a response based on an understanding of a mathematical concept connecting the two numbers, namely, place value.

- "Eighteen" is, of course, the hoped for mathematical response, but its occurrence depends not only on an understanding of the mathematically defined words for numbers but also on a mathematical interpretation of the ordinary language word *difference*.

Consider now the same question in its symbolic form: $20 - 2 =$. In the context of the ordinary primary classroom, where base ten is usually tacitly assumed, this question unambiguously requires the response "18."

The potential for ambiguity lies in more than just the choice of vocabulary, however. Were this not so, confusion or lack of comprehension could be avoided through the use of technical, mathematical terms. It would suffice to say, "What is twenty minus two?" a question that mirrors the symbolic form with verbal language. The problem lies in how children acquire the concept of minus in the first place.

We need to pause and consider how the *understanding* of mathematics comes about. The classroom environment exposes pupils to a combination of practical and aural experiences. In the particular case

of subtraction, pupils often work with physical objects, performing actions that involve "taking away" a number of items from a larger collection of items and also comparing the relative numerosity of groups of items by focusing on the "difference between" them. These activities occur within an environment of verbal language, and one could contend that that language has to be, at least initially, the language of the child and not the language of the mathematician. Children make sense of a concept and construct their own meaning through a combination of personal experience and cultural tradition. The more precise, mathematical language connecting the verbal with the symbolic can be offered only at a later stage. A link between semantics (the sense of the language) and semiotics (the symbolism) has to be created. For many children, the image of a mathematical concept will always be underpinned with the original everyday language that gave it birth. Ordinary language most likely comes first. It contains the initial images, and mathematical language can be superimposed only when some degree of understanding already exists.

That said, where does the problem lie? Surely children will simply replace the ordinary language they have been loosely using with the given specific mathematical terminology, will they not? It is true that many of them do appear to do so, and much of the time this replacement does not cause obvious problems, although we must remember that the only knowledge we can have of the pupils' understandings is gained from *our* interpretation of *their* communication to us through symbol or word. The growth of mathematical understanding occurs through a process of folding back to earlier images to give insight to the building of new, more powerful ideas (Pirie and Kieren 1994). Difficulties arise when the *language* of the original image inhibits the growth of understanding, that is, when the dominant ordinary language carries only a partial, incomplete image for the concept.

Consider the example of addition, a concept arrived at through, among other things, the actions of putting groups of items together and considering the new, whole group. These actions can be verbalized as "Six and three make nine" and symbolized as $6 + 3 = 9$. The words *add* and *equals* may be introduced as terminology to replace the *and* and *makes*, but the meaning attached to the new words is that derived from the ordinary language used. Not until much later, usually with the introduction of algebra, does the asymmetry of the verbal expression cause problems. Both $6 + 3 = x$ and "six and three make something" can be solved without difficulty. But for a significant number of children, $x = 6 + 3$ has no solution, since one cannot say, with meaning, "something makes six and three." It

is to this ordinary language reading of the symbolic expression that they turn when trying to interpret the symbolically written statement. For them, the result of an operation comes after the action, not before it. The ideas embedded in the original ordinary language impede the mathematical generalization of the notion of equations.

In many instances, the problems of connecting the ordinary, the mathematical, and the symbolic language occur from the fact that learning takes place through actions (physical or mental) performed over time and described in terms of active verbs, such as *make*, which here gives a definite image of left-to-right movement when "reading" the symbolic representation. *Take away* is another such active verb, which can frequently lead to the notion, often in fact actually voiced by many teachers, that "you cannot take a larger number from a smaller one." Physically, that is true, but that language later makes acquiring the concept of negative numbers harder to achieve.

An alternative to physical, time-dependent action as a means of introducing mathematical concepts are metaphors used by teachers to create images for their pupils. Of course, in the first instance these metaphors are presented in ordinary language. A common way to introduce solving linear equations is through the notion of a balance. Students are initially encouraged to think of solving the equations as a process of removing things from each side in such a way as to preserve the equilibrium of the balance. "Taking the same weight off each scale pan will not alter the balance" is a powerful image to use.

Jody and May had used and appeared to comprehend this notion of taking the same away from each side to leave the balance undisturbed and were writing such solutions as

$$\overset{0 \quad 5}{5x + \cancel{4} = \cancel{9}}$$

and saying, "Take four off each side." When faced with the question $5x - 4 = 2x - 9$, they responded, "Take four off each side" while writing

$$\overset{0}{5x - \cancel{4}} = 2x - 9 - 4.$$

They then said, "Take two x off each side" while writing $3x - 0 = -13$. They checked by substitution, found that the answer was wrong, and then spent some time and frustration repeating the calculation again and again, the whole while talking about "weighing the same" and "taking off weights." They were completely constrained by the images created by their language and unable to see what to do even when they

realized that their error was at the stage of taking off the 4. They used phrases like "You can't take it off, it isn't really there" (Jody) and "You can't do it by weighing" (May)—which is, of course, correct!

The image of equations as scales no longer holds literally for images for negative numbers. What does a weight of minus 4 kg look like? Obviously one could distort the image of "balance" a little here to explain the new situation, but the point is that for Jody and May, their language throughout the lessons indicated that their thinking rested not so much on the notion of balance as on the language of "equal physical weights." The language of the image encouraged a focus on numbers as concrete items with physical weight, which can be physically removed; this understanding had been strongly reinforced by its successful application over several lessons.

The move from ordinary to mathematical language is further made problematic by the retrospective quality of any new, imposed mathematical vocabulary. Notions that themselves do not at first appear identical or even similar, such as "the removal of objects from a collection" (take away) and "the comparison of two collections" (difference between), are labeled with the same mathematical word (*minus*) because the result of the two actions is the same; thus from the mathematician's viewpoint, both are manifestations of the same mathematical concept. Further evidence of this necessary elision of meaning occurs when we deal with negative numbers and subtraction. For the integer –3 (verbally rendered as "negative three"), the "–" sign has an adjectival role, whereas in the statement 2 – 3, the role of the "–" sign is as an operation, an action, a verb in the word form. For the mathematician, the images and associated language provide metaphors for the mathematics; for the learner, frequently they *are* the mathematics.

The concept of division offers another vivid illustration of the encapsulation of two completely different images within one mathematical notion. Although either image is sufficiently powerful to allow an understanding in terms of integers, it is in fact essential that children have both images if they are to move to a comprehension of division with fractions.

Consider 12 ÷ 3, verbalized as "twelve divided by three." This can be interpreted as "twelve pizzas shared among three people" and diagrammatically represented as follows:

oooo   oooo   oooo

Here, the answer lies in the number in each group. Alternatively, it can be interpreted as "how many groups of three pizzas can be made from twelve pizzas?" and diagrammatically represented as follows:

Part 3: Different Styles and Patterns of Communication in the Mathematics Classroom

Contents

Part 4: Problems of Communication in Specific Domains of Mathematics

ooo   ooo   ooo   ooo

Here, the answer lies in the number of groups.

The learner is left with two totally different images, both in the visual and the mental sense. The ordinary language of either "sharing" or "grouping" can be used to describe actual actions or to conjure a metaphor that gives meaning to the mathematical language of "division," *but* these actions and metaphors are different.

Next consider "a half divided by 3." Translating this to fit the previously given images, we see that "half a pizza shared among three people" makes sense, whereas "how many groups of three pizzas can be made from half a pizza?" leads to the answer "none"! The second image is of no use here. However, now consider "12 divided by a half." Translating this to fit the images, we get "twelve pizzas shared between half a person," which is meaningless, whereas "how many half pizzas can be made from twelve pizzas?" is clearly a sensible question.

Once again the choice of ordinary language constricts the understanding of the mathematics.

FROM VERBAL MATHEMATICAL TO SYMBOLIC LANGUAGE

So far we have been looking at the move between ordinary and mathematical language, but even the move from verbal to symbolic language is not without its problems. Within the mathematical register, verbal and symbolic forms do not always match. Here are two trivial, but nonetheless real, examples: "subtract two from three" is not written "– 2 3" but "3 – 2"; we say "add two, three, four, and five," but we write "2 + 3 + 4 + 5," not "+ 2, 3, 4 + 5." The real problem here, however, lies with the introduction of this third form of communication—mathematical symbolism. A verbal rendering of mathematics is mathematics mediated through language, be it ordinary or mathematical, and therefore through the life experiences of the reader or hearer. In its symbolic form it is something more. A mathematical symbol has abstract meaning that may be considered, in some fashion, absolute. It has no subjectivity. For the mathematician, the power of the symbol lies in its ability to be unambiguous and at the same time to encompass a range of illustrative images.

Mathematical symbolism in some way *is* the mathematics, independent of any experiences of the reader, and yet it is open to interpretation only through the medium of verbal language, which relates the mathematics to the reader's previously comprehended metaphors. It is here that we begin to see a rift between meaning and understanding that needs exploration if we are to better comprehend the role of communication in mathematics. Consider what meaning the symbol "3/4" communicates. How is this meaning verbalized? "Three quarters" (a number)? "Three items out of four" or "One item divided into four pieces, take three" (both quantities but with different visual images)? "Three over four" or "three slash four" (a way of writing)? "Three divided by four" (a process)? The symbolic representation is intended to encapsulate all the mathematical meanings, but how can pupils and teachers know which images are triggered for one another by the symbolism? Conversely, how can we know what meanings are transferred from the current understanding of the mathematics to the personal writing done in symbolic form?

One of the strengths of symbolism is its brevity, but this in itself can be the root cause of misunderstandings. Pupils trying to move between the verbal and symbolic writing strive for the same brevity and lose the meaning.

Mark was working on evaluating the equation $y = x^2 - 1$ for various values of x. Talking as he wrote, he said:

So y equals x squared minus one (writing $y = x^2 - 1$).

At x equals minus one, we get minus one times minus one, minus one (writing $y = -1\ -1\ -1$ as if he were writing $ab - c$).

Two minuses make a plus, er … plus, er … plus times minus makes minus, so minus one minus one minus one is minus three (writing $y = -3$).

In all probability the mathematicians' shorthand way of writing a times b as ab has added to the error associated with the verbal "rule" for multiplying negative numbers. The teacher spent some time with Mark, trying to explain the problem by changing the language to "negative one squared subtract one" (writing $[-1 \times -1] - 1$), but several weeks later Mark was still confidently saying "minus three times minus two is … is … minus minus is plus so is … five" as he wrote $-3 - 2 = 5$. The teacher's attempt to introduce a more helpful verbalization by using different mathematical words was confounded by Mark's knowledge that *minus*, *subtract*, and *negative* were interchangeable words for the same symbol, namely, "–." It is likely that he was not even listening with care to the language of the teacher, since for him it did not appear to conflict with his own verbalization.

On a more positive note, the following transcript of a conversation among Katie, Carol, and Alison as they worked on the area of trape-

Colin: ... and turn it off ...

Ali: ... here that's like that, there's two leads going off that way because you can stack 'em up, can't you? And they just go right up the side, they don't have to.

Colin: Yes it does.

Ben: It does.

Ali: They stack up like that.

Colin: Oh, that's right.

SP: Well hang on, I'm confused. We don't have the circuit in front of us. What is going in where, and what is being connected?

Ali: These little leads. (*Indicates imagined objects*)

Colin: These leads and wires.

SP: OK, what are the wires to and from? I haven't understood what you've connected to it.

Ben: Well, there's one from this light sensor...

Colin: ...to the buzzer.

Ali: When you push it, the buzzer will go off.

Colin: And then you cover it, don't you?

Ben: You want to cover this so it's dark and that will go off.

Ali: That's a direct link, yeah.

Here the pupils have a highly visual and not at all well verbalized understanding of their ways of thinking. It appears that even four months after the lessons had occurred they all share the same understanding and do not need explicit language to express their meanings. The only person in the dark is the interviewer! The lesson to be heeded here is that one must not confuse an inability to talk lucidly about mathematics with an inability to do or even understand mathematics. In the example given, an assumed common understanding of the mathematics and the tasks had obviated the need to develop unambiguous language. For these boys, a lack of language did not imply a lack of understanding.

The next extract looks at two boys who are not without language to discuss their mathematics, but whose unverbalized but shared meaning does not accord with that of the teacher. The pupils were working together on linear equations, making tables of values and plotting graphs. One of the equations given on the worksheet was

$$y = 2x + 3.$$

Kevin: You fill it. (*Referring to their table*)

       x ... one, x squared ... two, x squared plus three ... five.

       x ... two, x squared ... four, x squared plus three ... seven.

       x ... three, x squared ... six, x squared plus three ... nine.

       Four ... Wait, we need some minuses ... er ... do minus three.

Dave: Easy-peasy ...

       Minus three: x squared ... minus six, ... minus three

       Minus two: x squared ... minus two, ... no ... minus four, ... minus one

       Minus one: x squared ... minus two, ... plus one

At this point the teacher passed by and hearing "x squared," stopped.

Teacher: Hang on. Go back a bit. You said, "If x is three then x squared is six?"

Kevin: Yeah.

Teacher: OK. Hang on a minute....

He then proceeded to draw a 3×3 square, divide it into unit squares, and count them.

Teacher: So it's nine, isn't it?

Kevin: What?

The teacher repeated the explanation and demonstration with 4×4 and 5×5 squares.

Kevin: Oh, yeah. (*Doubtfully*)

Teacher: Can you carry on now?

Kevin: Yeah.

The teacher moved away.

Kevin: What's he on about? I know all that stuff about areas and counting things.

Dave: Yeah, what we got here's lines.

Kevin: Just ignore him. Your turn. You do x squared minus three. (*Writing $2x - 3$*)

This transcript illustrates again a need on the part of teachers to separate the use of language from indications of understanding. Here the teacher has heard a misuse of language—*squared* for *twice*—but interpreted their problem as one of misunderstanding the notion of "squared." In this instance, the subsequent discussion neither enhanced nor inhibited the understanding of the boys, since they chose to ignore the teacher's input. The pupils possessed a shared, correct understanding of the mathematics that was not challenged by the language they used, even though the language was incorrect. A more appropriate intervention by the teacher might have been to simply remark that $2x$ is read "two x" and that x^2 is read "x squared." The problem was one of language—the correct relationship between verbal and symbolic language—and not one of mathematical understanding.

There is, of course, always the danger that ideas that are assumed to be shared but remain unverbalized may not hold the same meanings for all participants. Frequently this can occur between teachers and pupils and, when spotted, is commented on with such teacher language as "I assumed they knew that...." Such situations can also, however, occur between pupils themselves and, passing undetected, lead to misunderstandings that are not evident to anyone. Of three pupils working on a problem concerning $x^2 < z < y^2$ and talking about "the numbers between," two were using the word *between* to be related to those numbers that are physically between, say, 4 and 9 on a number line, that is, the four numbers 5, 6, 7, 8. The third, however, was giving the word its mathematical sense, meaning "difference between," that is, in this example, the number 5 ($9 - 4 = 5$). All three were unaware of this lack of common meaning, and confusion ensued (Pirie 1991).

The use of quasi-mathematical language will, I suspect, continue to cause controversy between mathematical linguistic purists and more pragmatic teachers. It is a practice that is to be found among all communicators of mathematics, with a greater or lesser degree of acceptability. It generally arises when no mathematical language is readily available, when the language is too sophisticated for the learner, or when a metaphorical image is taken too literally.

This first example comes from a group of boys who were using language that was ordinary and had been appropriate in an earlier context. Since then, no obvious mathematical term being at hand, the boys had, by unspoken consent, given a specific mathematical meaning to the word *moves*. In a previous lesson these pupils had worked together on an investigation called "Frogs," which involved counting the number of times that counters were moved to achieve a prescribed rearrangement. In the following extract, however, the pupils are counting how many ways they can arrange a given number of square tiles.

Len: There are 7 moves for 4.

Andy: How many different moves did you get when you got to 5?

Len: You should have wrote yours down.

Andy: (*Dismissively*) I have! How many number of moves—how many number of moves did you get?

Chas: For 2 it's 1. For 3 it's 3.

How they maneuver the tiles to produce the arrangements is irrelevant and not being counted or attended to in any way, yet well into the lesson they are still using the language of *moves* to mean *arrangements* without any confusion to themselves. They have implicitly, but nowhere explicitly, created a shared understanding that, say, "in mathematics lessons we will use the word *moves* to mean anything that we have been asked to count." It is interesting to note that several weeks later, these same pupils were still using the word *moves* in relation to number patterns even when the patterns being investigated were presented purely symbolically. Given the sequence 1, 4, 9, 16, ..., they set about calculating "the fifth move in the pattern."

What the preceding example, and indeed that of Kevin and Dave, illustrates is a need for vigilance when pupil-pupil discussion is encountered in the classroom. On the one hand, pupils can quickly develop a shared quasi-mathematical language, carrying meaning for the users, that is unorthodox or incompatible with accepted mathematical language. On the other hand, however, this quasi-mathematical language leads to no difficulties in understanding and can, in fact, often enhance understanding by forming a language-linked image that is of personal relevance to the learner. Textbook writers themselves are not above creating such quasi-mathematical language. The angles known to mathematicians traditionally as "corresponding" are termed "F" angles by the SMP (School Mathematics Project) textbook series; one can see how this image of the angles trapped by the parallel arms of the F could be a visual aid to remembering which angles have the same values. (See fig. 1.5.)

In one classroom, however, they became known as "parallel angles" to the majority of pupils. Invented by two of the pupils in the course of working together ("they are the same in a different place, like parallel lines are"), the name appealed to the teacher as "better than SMP's 'F' angles. Children don't recognize 'F' angles when they are upside down, and they do see them when the lines are not necessarily parallel. Half of them don't know what *correspond* means anyway. The kids understand why they are the same so I'm happy to call them what they like." Unorthodox maybe, but how many of his children carry a better understanding with their quasi-mathematical term?

Angles and Parallel Lines: F-Angles

Any pair of parallel lines make
an F-shape with a line
that crosses them.

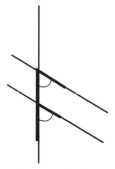

The marked angles are equal.
We shall call them F-angles.

Find the angles marked by letters in the diagram below.
Look for F-shapes. In the first one,
an F is picked out for you.

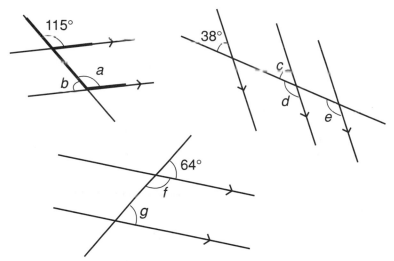

Source: School Mathematics Project, *Angle Relationships* (SMP 11–16 2e), pp. 4–5.
Reprinted with permission of Cambridge University Press, all right reserved.

Fig. 1.5

Other examples of such seemingly apposite, invented language
include the obvious use of *length of a circle* for its diameter and the
ingenious coining of *divising* [*sic*] for finding a common denominator
(a mixture of notions involving division and devising an appropriate
number?). For pupils in one class, the verb *to Y* a pizza meant to divide

it into thirds (fig. 1.6a), whereas for those in a different class, to get one third of a circle, they *cheesed* it (fig. 1.6b).

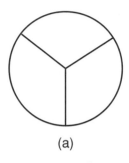
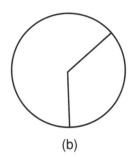

(a)   (b)

Fig. 1.6

These shared meanings arise often and usually without explicit explanation because, when talking with one another, pupils are most likely to revert to ordinary language and seem happy to create spontaneously these quasi-mathematical vocabularies as and when they are needed.

Although nowhere in this chapter do I begin to consider the language problems that arise for pupils whose mother tongue is not used in the classroom, by the teacher, or in the textbook, I intend to finish with a transcript of Abu working in an SMP classroom. Along with many other textbook writers, SMP authors have given a great deal of thought to the issue of language and image in their series of books and booklets for students from 11 to 16 years of age. Abu was at a school with a very high intake of children for whom English was not the first language, and the SMP 11–16 series had been deliberately chosen by the mathematics department with this situation in mind. In Abu's class, the children were allowed to work together and discuss their work as a means of overcoming any language problems that presented themselves. They worked their way through the collection of booklets, sometimes guided by the teacher and sometimes given the freedom to choose their own. In this extract, Abu had just returned from talking to the teacher, and I stopped him on his way to change his booklet.

SP: Before you start the next book, tell me what you've just finished. Did you get them right?

Abu: Yeah. I just checked them answers with Mr. M. and he said to choose another one.

SP: What was the booklet about?

Abu: It was 2b (see fig. 1.7).

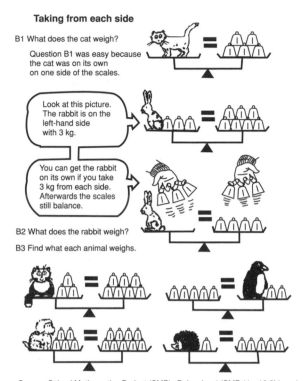

Fig. 1.7. Taking from each side

SP: And what did you have to do?

Abu: The first bit's boring. You drawed all these pictures of bells and animals and things and guess the weight of the animals and then you can write things with balanckies and things.

SP: (*Having no idea what he was talking about*) OK. Show me some of your drawings. Let me see what you mean.... Draw me something.

(Abu produces the drawing in fig. 1.8.)

SP: ... So how do you work out the weight of the rabbit?

Abu: Well you sort of guess ... you know ... you must guess.

(Since he had got them all right, SP assumed that he was not "just guessing" but was unable to verbalize his understanding, and so SP tried to push him a little.)

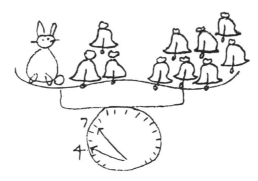

Fig. 1.8. Abu's drawing

SP: OK, was the whole book like that?

Abu: No. Mustaph [another pupil at his table] said I could stop drawing, and I can't draw very good.

SP: What did you do when you stopped drawing?

Abu: I drew balanckies and guessed what *they* weighed.

SP: Can you show me?

Abu: So … well, an easy one…. (*He wrote ? = 3.*)

SP: Uhuh.

Abu: The balancky is 3.

SP: (*Still mystified*) Go on. Show me a hard one.

Abu: Errr … (*Eyes shut*) balancky and balancky and 3 is 6. No, 7. The balancky has to be 2.

SP: Can you write that down?

Abu: (*Writes ? + ? + 3 = 7   ? = 2*)

SP: So if I wrote: (*Writes ? + ? + ? + 2 = ? + 10*) could you do it?

Abu: (*Eyes shut*) Err…. (*Opens eyes and writes ? = 4*) The balancky is 4.

His algebraic thinking is quite correct, of course, but where have *balanckies* come from? The booklet is actually called *Balancing 1*—read by Abu as "Balancking 1" and conveying no meaning to guide his understanding.

Never having seen a scale pan of the old-fashioned kind illustrated in the booklet, he had no notion that the image of a balance could be helpful. The stylized drawings of weights he saw as bells! Abu's lack of language prevented him from accessing the image—that of balancing sides to an equation—that the text is trying to convey. A later page in the booklet, however, reveals some meaning for him (see fig. 1.9). Once

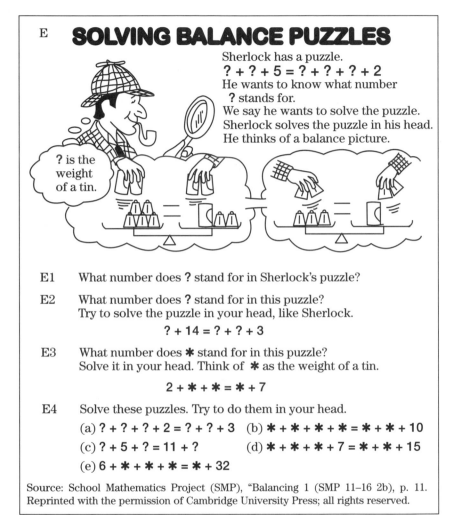

Fig. 1.9. Solving balance puzzles

faced with an "algebraic" notation—that of question marks—he finds a meaning for the unknown language, calls the new symbol "?" a *balancky*, and clearly has no trouble with the algebraic concept. It was amusing to notice that a week later all the pupils at Abu's table were calling a question mark a *balancky*.

A postscript to this episode came three weeks later when Abu came running over to see the interviewer and said:

Abu: Miss, I'm on algebra now!

SP: Wow! Show me what you are doing.

Abu: Did you know all them balanckies we wrote was algebra? And there's an easy way. You just write a balancky in bracket [parenthesis] for lots of them, like (*Writes* 8(?)) is really 8 of them. Clever, i'n'it!

Clearly, here was a lad for whom symbolic language was a liberation from his shaky ordinary language.

FINAL REMARKS

The main message of this chapter is that we cannot assume from children's ability to perform mathematical tasks that they possess the language to talk about their understanding, nor, probably more important, can we infer that incoherent language used in an attempt to communicate the mathematics being done reflects a lack of understanding. There *are* connections among pupils, their language, and their mathematical understanding, but we, as observers and teachers, must not make superficial judgments. All we can ever work from when trying to access the understanding being constructed by pupils is their language—of whatever form, verbal or symbolic. It is through their language that they express their current mathematical understanding, and our understanding of them is limited by the ways in which they try to express this understanding. In addition, our interpretations are also shaped by *our own* construction of the meanings they express. We need consciously to work hard to uncover the mathematical reality that each individual pupil creates—and to bear in mind that this reality can be revealed to us only through the mediation of language and supposedly shared meanings.

REFERENCES

Pirie, S. E. B. "Peer Discussion in the Context of Mathematical Problem Solving." In *Language in Mathematical Education*, edited by K. Durkin and B. Shire, pp. 141–61. Milton Keynes: Open University Press, 1991.

Pirie, S. E. B., and T. E. Kieren. "Creating Constructivist Environments and Constructing Creative Mathematics." *Educational Studies in Mathematics* 23 (1992): 505–28.

———. "Growth in Mathematical Understanding: How Can We Characterise It and How Can We Represent It?" In *Learning Mathematics: Constructivist and Interactionist Theories of Mathematical Development*, edited by P. Cobb, pp. 61–86. Dordrecht, Netherlands: Kluwer Academic Press, 1994.

School Mathematics Project (SMP). *Angle Relationships* (SMP 11–16 2e). Cambridge: Cambridge University Press, 1984a.

————. *Balancing 1* (SMP 11–16 2b). Cambridge: Cambridge University Press, 1984b.

2

Three Epistemologies, Three Views of Classroom Communication: Constructivism, Sociocultural Approaches, Interactionism

Anna Sierpinska

Concordia University, Montreal, Quebec

ANGUAGE and communication in the mathematics classroom have become much discussed topics recently. This is not to say that mathematics educators have not occupied themselves with the issues before; they have, but the accent was more on language than on communication, and the discussions and research had more the flavor of "normal science" (in the sense of Kuhn). The present-day fuss over communication, language, and discourse in the classroom takes place at the point in the history of mathematics education where certain long-held views and perspectives are questioned and new ones are noticed and promoted. The social and cultural aspects and the emergent character of the mathematical meanings jointly constructed by the teacher and the students in the classroom are getting the limelight previously enjoyed by the cognitive processes of individual students. Mathematics education is traversing a period of redefining its basic categories. Language in mathematics education has always been an issue, but now the attention has shifted from the study of texts to the study of language in action—its use in different contexts and as a part of social practices; in brief, the focus has moved from language to discourse. Some time ago, one of the main questions in mathematics was, How can we best communicate such and such notion *to* the students? More recently, it became, Does communication [of one's thoughts] enhance cognition? Today, attention revolves around the processes of communication *among* students and communication *with* students and the question of the emergence of taken-as-shared meanings through communication in classroom cultures. Communication is not understood as just a necessary means of education; education is identified with communication:

> [E]ducation is best understood as a communicative process that consists largely in the growth of shared mental contexts and terms of reference through which the various discourses of education (the various "subjects" and their associated academic abilities) come to be intelligible to those who use them. (Edwards and Mercer 1993, p. 63)

The different understandings that have been attached to language and communication in mathematics education are not isolated ideas; they have grown on and within certain larger theoretical systems and epistemological perspectives. The aim of this chapter is to sketch the evolution of these understandings within three broad theoretical approaches: the constructivist or Piagetian, the sociocultural or Vygotskian, and the interactionist or Brunerian. All of these approaches are presented, represented, or discussed in the chapters that follow.

A CONSTRUCTIVIST APPROACH TO LANGUAGE AND COMMUNICATION: STUDENTS TALKING, TEACHERS LISTENING

Communication as a Theoretical Problem

As noted by several authors (Cobb 1994; Gergen 1995; Richards 1995), from a constructivist point of view, communication is a problem, in the sense that it is hard to explain how it is at all possible. In chapter 6 of this volume, Kanes mentions the difficulty of using constructivism to explain the phenomenon of communication of meaning and discusses several alternative accounts of the formation of meaning in language.

The constructivist view of the solo mind trying to make sense of its experience with and in the environment replaces the concept of communication with a more cautious one of tentative interpretations of other people's words and sentences.

> At best we may come to the conclusion that our interpretation of [other people's] words and sentences seems compatible with the

model of their thinking and acting that we have built in the course of our interactions with them. (von Glasersfeld 1988)

Communication is thus seen from the participating individual's perspective, and, in fact, it remains in the background of research and reflection. The focus is on the individual, on the cognitive aspects, and on the development of the psychological self and its mental structures.

Language as an Expression of Thought

"Language is molded on the habits of thought" (Piaget 1959, p. 79). Piaget claimed that the way children speak and talk reflects their immature thinking. Egocentric thought produces egocentric speech.

The difference between egocentric speech and social speech is vividly depicted in the following excerpt:

> We shall quickly realize the full importance of egocentrism if we consider a certain familiar experience of daily life. We are looking, say, for the solution of some problem, when suddenly everything seems quite clear; we have understood, and we experience that sui generis feeling of intellectual satisfaction. But as soon as we try to explain to others what it is we have understood, difficulties come thick and fast. These difficulties do not arise merely because of the effort of attention needed to hold in a single grasp the links in the chain of argument; they are attributable also to our judging faculty itself. Conclusions which we deemed positive no longer seem so; between certain propositions whole series of intermediate links are now seen to be lacking in order to fill the gaps of which we were previously not even conscious; arguments which seemed convincing because they were connected with some schema of visual imagery or based on some sort of analogy, lose all their potency from the moment we feel the need to appeal to these schemas, and find that they are incommunicable; doubt is cast on propositions connected with judgments of value, as soon as we realize the personal nature of such judgments. (Piaget 1959, pp. 45–46)

Thus, to become communicable, thinking must undergo an evolution. A substantial developmental step must be taken between egocentric speech and social speech. Each reflects a different type of thinking.

Language as a Symptom of Thought: Teachers Should Listen to Students

Language, understood as a symptom of thought, becomes the teacher's or the researcher's instrument for building models of the child's or the student's thinking. By studying children's spontaneous language, Piaget expected to obtain some evidence about children's ways of thinking and reasoning. Constructivist mathematics educators advocate that teachers listen to children, construct models of their ways of thinking, and build their teaching activities on these models. Von Glasersfeld makes a sharp distinction between "training" and "teaching" in this respect:

> Whereas the trainer focuses only on the trainee's performance, the teacher must be concerned with what goes on in the student's head. The teacher must listen to the student, interpret what the student does and says, and try to build up a "model" of the student's conceptual structures. This is, of course, a fallible enterprise. But, without it, any attempt to change the student's conceptual structures can be no more than a hit or miss affair. In the endeavor to arrive at a viable model of the student's thinking, it is important to consider that whatever a student does or says in the context of solving a problem is what, at the moment, makes sense to the student. It may seem to make no sense to the teacher, but unless the teacher can elicit an explanation or generate a hypothesis as to how the student has arrived at the answer, the chances of modifying the student's conceptual structures are minimal. (von Glasersfeld 1995, p. 15)

In the construction of a viable model of a student's thinking, however, it is necessary to take the Piagetian assumption that "speech is molded on habits of thought" with a grain of salt. Research on children's mathematical behavior suggests that one has to be very careful in interpreting *what* they say on the basis of *how* they say it. Incoherent talk need not be an expression of incoherent thought, as Pirie has pointed out in chapter 1 of this volume.

Communication as an (Impossible) Transmission: "Concepts Cannot Simply Be Transferred from Teachers to Students—They Have to Be Conceived"

If speech is an encoded thought, then communication means transmission of thoughts mediated by language. The hearer decodes the utterance of the speaker to reach to the thought. There is no warranty, however, that decoding is an inverse operation to encoding, hence the possibility of error and, thereby, the theoretical impossibility of the transmission of thoughts (for an in-depth discussion of constructivism's difficulty in explaining intersubjectivity, see Lerman [1996]).

The notion of communication is thus a problem for constructivism because it has been linked with the transmission of thoughts. One proposed solution is to replace communication by compatibility between the *intra*individual coordination of actions and their *inter*individual coordination:

> [L]ogic is before all else the expression of the general coordination of actions, and ... this general coordination of the actions necessarily includes a social dimension, since the inter-individual coordination of actions and their intra-individual coordination constitute a single and identical process, the individual's operations all being socialized, and co-operation in its strict sense consisting in a pooling of each individual's operations. (Piaget 1972, p. 71)

This excerpt paints the picture of communication as a direct sum of the work of two solitary minds.

In other approaches, transmission may not be an issue, since verbal interaction and communication are seen in a more holistic way as possessing emergent properties, transcending the individual inputs.

The claim about the impossibility of verbal transmission has become a slogan of constructivism in mathematics education (sometimes it is reduced to this slogan, as deplored by von Glasersfeld [1994]). Constructivists repeat that students cannot be taught by "telling" (Arcavi and Schoenfeld 1992, p. 323) because "knowledge is actively constructed by the cognizing subject, not passively received from the environment" (Kilpatrick 1987, p. 7). This assumption comes to the forefront in the more concise presentations of the basic tenets of constructivism. For example:

> We begin with the assumption that students actively construct meaning. They are not tabula rasa upon which teachers "write knowledge." Each student makes sense of the world in terms of the understandings of the world that he or she brings to it. These understandings or models of the world are constantly being revised, and we are never in a final state. Thus we are in general accord with the precepts of what has become known as constructivism. (Lesh and Kelly 1994, p. 277)

Constructivists sometimes revise themselves, however, and say only that knowledge that can be transmitted verbally is not very valuable. A student can be taught a formula by being told it, and he or she can then apply it to solve a number of typical exercises. This experience will not, however, prepare the student to deal with problems that are not so typical.

> The solving of problems that are not precisely those presented in the preceding course of instruction requires conceptual understanding, not

only of certain building blocks but also of a variety of relationships that can be posited between them. Only the student who has built up such a repertoire has a chance of success when faced with novel problems. Concepts cannot simply be transferred from teachers to students—they have to be conceived. (von Glasersfeld 1995, p. 5)

What has come to be understood, in mathematics education, as a constructivist style of teaching? Which pedagogical interventions harmonize with the theory and which do not? Many authors have reflected on this question (see, for example, Davis, Maher, and Noddings [1990], and especially Confrey's paper therein; also Larochelle and Bednarz 1994; Steffe and Gale 1995).

In some accounts of the constructivist classroom, the merits of communication in social interactions are put forward as if communication were not a problem in the theory. They bear witness to a shift toward a more sociological perspective in their authors and a departure from radical constructivism.

From [the constructivist] perspective, teaching is viewed as more than providing information and checking to see if it has been acquired by students. Instead, teaching becomes a matter of creating situations in which children actively participate in scientific, mathematical, or literary activities that enable them to make their own individual constructions. To teach well from this perspective, teachers will need opportunities in which they can learn about their students' constructions. This can be accomplished by creating settings that encourage children's sensorimotor and mental activity and providing social situations in which communication can take place. Some examples of such social arrangements are whole-class discussions of scientific experiments, small-group collaborative problem solving in mathematics, and written drafts shared with others in the course of composition writing. (Wood 1995, p. 337)

Other authors, trying to remain faithful to the foundation of constructivism, struggle with the problem of communication.

Teaching, in general, requires that many content- and pedagogically-grounded decisions be made while interacting with students. Teaching from a constructivist perspective imposes additional constraints. The selection among many possible courses of action should be deeply respectful of students' previous knowledge and construction processes, in essence "following the student's lead" where possible—perhaps pursuing nonstandard but ultimately profitable mathematical directions. Yet it should also make life easier for the student where possible, for example by helping students extricate themselves from messy situations or avoiding major pitfalls. The teacher must make numerous split-second decisions. When is an intervention, of what type, appropriate? Is "telling" always undesirable, or are there times when it is useful? And what does one tell? To what extent should students' idiosyncratic constructions be encouraged, supported, and pursued? (Imagine a world in which every student developed his or her own good sense of whole number operations, but did so in a personal idiosyncratic language that made communication with others difficult if not impossible!) (Arcavi and Schoenfeld 1992, p. 323)

Let us have another look at what the constructivist theory might imply for the mathematics classroom communication.

Can Communication Enhance the Development of Thought?

Constructivist mathematics educators often stress the positive role of social interaction and communication in the mental development of students. In what sense are they consistent with Piaget's theory, and in what sense are they not?

According to Piaget, mental development is seen as dependent on biological-neural maturation of the brain functions (Piaget 1972, p. 36). Hence, he claims, the interactions with the environment cannot in any way cause the development. In particular, schooling can perhaps speed up the development a little, but schooling cannot make it happen. A constructivist teacher will thus follow the development of the child rather than precede it.

> [T]he development of intelligence ... is dependent upon natural or spontaneous processes, in the sense that they may be utilized and accelerated by education at home or in school but that they are not derived from that education and, on the contrary, constitute the preliminary and necessary condition of efficacy in any form of instruction.... [I]ntellectual operations constitute the expression of neural coordinations that develop as a function of organic maturation alone. (Piaget 1972, p. 36)

Developmental maturity is thus a prerequisite condition for education and communication to be effective. It is the task of psychology to study the conditions of the success of communication and education.

> Apart from the factors of maturation and experiment, the acquisition of knowledge naturally depends upon educational or social communications (linguistics, etc.), and for a long while it was solely to this process that the traditional school confined its attention. Psychology in no way wishes to neglect such communication but sets itself to study questions that affect it and that may have been supposed to be long since resolved: does the success of such communication depend solely upon the

quality of the presentation made by the adult himself of what he desires to inculcate in the child, or does it presuppose in the latter the presence of instruments of assimilation whose absence will prevent all comprehension? (Piaget 1972, p. 39)

These conditions include the development of the mechanisms of assimilation and accommodation and the construction of the operational structures of the mind.

Thus the belief of mathematics educators in the benefits of discussion in the mathematics classroom is *theoretically* justified with respect to children whose thinking is developmentally mature enough to be communicable. Organizing discussions with younger children is not going to speed up their natural development.

The fact of being or not being communicable is not an attribute that can be added to thought from the outside, but is a constitutive feature of profound significance for the shape and structure which reasoning may assume. (Piaget 1959, p. 48)

At a later age, indeed, discussions can create opportunities for the children to practice, or put to work, their potential to think for the purposes of communication. It can incite them to look for convincing arguments and systematic frames of reference, to synthesize whole chains of reasoning, and to perceive different possible points of view and rationales for value judgments.

According to Piaget, when the child starts actually to discuss his or her ideas with others, it is *because* his or her thinking is more geared toward comparing different points of view, and not the other way around. But, of course, once the child is ready, "the most important occasions for accommodation arise in social interaction" (von Glasersfeld 1995, p. 11, see also Piaget [1976, p. 137]).

Even then, however, we may be quite disappointed, in mathematics education, in the effects of discussion on students' mathematical understanding (see chapter 11 by Stacey and Gooding in this volume). The reason can be that in the practice of teaching mathematics, we expect students not to resort to *any* kind of justifications and achieve *any* kind of accommodation but some very specific ones: mathematical, satisfying certain definite standards, internally consistent, and compatible with mathematical theory.

In chapter 10, Curcio and Artzt bring to our awareness the fact that quite a few conditions have to be satisfied for a small-group discussion to contribute to the growth of mathematical understanding in the participants and to actual progress toward the solution of the problem with which the group is dealing. Students must engage in a variety of cognitive and metacognitive activities, such as understanding, exploring, planning a solution strategy, implementing it, and verifying the results and hypotheses. Examples of small-group discussions among preservice elementary school teachers, given by Civil in chapter 12, confirm this need and make it clear that the existence of disagreement in a group does not automatically lead to a better understanding by the students or to their overcoming the obstacles. Other factors, such as beliefs about what counts as mathematics, the person's trust in his or her mathematical abilities, and the group's previous mathematical experiences, have to be taken into account.

A "Constructivist Style" in Mathematics Teaching: A Child-Centered Pedagogy

Communication in teaching mathematics is of an especially delicate nature because mathematics is not about things that can be shown, nor is it even about relations between such things. Most of the time it is about relations between relations. It is not about things that are transformed nor about the transformations, but about the mechanisms or rules of these transformations. Theoretical constructivism's problems with the notion of communication notwithstanding, the very nature of mathematics makes it necessary to teach this subject indirectly.

To illustrate a constructivist attitude toward the use of language and communication in teaching, let us consider the well-known Piagetian-style activity (Piaget 1972, p. 40; Donaldson 1978, p. 61; Edwards and Mercer 1993, p. 58) in which the teacher first shows the child that two sticks or rulers are congruent by superimposing one over the other and then puts the sticks before the child in the following position (fig. 2.1):

Fig. 2.1

The child says that the upper one is longer. What is the reaction of the constructivist teacher? There is no attempt to train the child to produce the right answer, nor is there an attempt to explain why the child is wrong. The teacher questions the child only to better understand what he or she thinks, asking why the child claims what he or she does. A very cautious teacher would not even challenge the child's arguments by saying, for example, "But, look, when I put them together, they are exactly the same." Rather, the teacher would only repeat the action of putting the two sticks together and then drawing them slightly apart.

On the basis of the child's explanations, the teacher may conclude that the child understands the term *longer* in an ordinal and not metric sense, as "reaching further" (Piaget 1972, p. 40). A researcher will be satisfied with such an explanation and lose interest in the child after the experiment. The pedagogue has some responsibility toward the child, so he or she will think about doing something to start the child thinking about *longer* in the more generally accepted metric sense.

What choices does the pedagogue have?

If the pedagogue is a Piagetian constructivist, he or she will refrain from verbal explanations because he or she believes that the source of understanding is in the individual's actions on physical or mental objects and their transformations in the aim of grasping "the mechanisms of those transformations as they function in connection with the transformative actions themselves" (ibid., p. 29). For him or her, "language is not sufficient to transmit a logic, and it is understood only thanks to logical instruments of assimilation whose origins lie much deeper, since they are dependent upon the general coordination of actions and of operations" (ibid., p. 40).

Nor will the constructivist teacher reach for the so-called visual methods, trying to show the student what it means for a stick to be longer than another stick by displaying sticks of various lengths in front of the student's eyes. The constructivist would thus trigger only a kind of figurative process, whereby mental images of the displayed objects would be created, but this copy of reality cannot be called knowledge. The source of knowledge is in the individual's actions on physical and mental objects (ibid., p. 72). Moreover, for the teacher, the metric notion of longer is of a logico-mathematical nature; logico-mathematical operations are abstractions not from objects but from transformations of objects and relations between transformations, and hence from things that are already of a mental and operational nature (ibid.).

For a constructivist, knowledge comes into being as a solution to a problem (Laborde 1994, p. 146). Thus, in our example, instead of using verbal explanations or visual presentations to help the student understand *longer* in a metric rather than ordinal sense, the teacher will try to create an opportunity for the student to engage in an activity in which the metric notion of longer would have an essential part to play, for example, an activity of fitting shelves to pairs of shelf brackets. The shelves at the child's disposal would be too short, so he or she would have to ask for *longer* shelves. To make it necessary for the child to use the word and not just to show what he or she wants by gestures, some constraints could be put on the situation; for example, the child would have to make the order by telephone or to a person hidden behind a curtain.

If the child in our example, after having gone through this and some other activities, still has not changed his or her way of understanding the relation of *longer*, the teacher will consider him or her still immature for the change and postpone further attempts until later.

A constructivist teacher will refrain from imposing any kind of technical language on students, especially on the younger ones. This teacher will refrain from giving explanations that are not responses to the students' explicit requests. He or she will see as most important the task to "guide the student into forming his own ideas and discovering mathematical relations and properties himself" (Piaget 1972, p. 48). It will also be seen as necessary "to study the mistakes made by the students and to see them as a means of understanding their mathematical thought and to train students in the practice of personal checking and autocorrection" (ibid.).

The students will be "interviewed" with the aim of eliciting and improving their understanding; they will rarely if at all be "interrogated" and asked to "recite" what they have learned. In any instance, "diagnostic evaluation" will be supplemented by "formative evaluation" (Bergeron, Herscovics, and Nantais 1985, pp. 13–19).

It appears that from the constructivist point of view, the teacher's language can become an obstacle to the learning of mathematics, whereas the child's own language is a prop. This is a child-centered pedagogy with asymmetric relations in communication. The child is talking and the teacher is listening and proposing adequate and challenging problem situations.

CRITICISMS OF THE PIAGETIAN THEORY

Much of Piaget's theory about children's judgment and reasoning has been criticized on the grounds that his analyses of the experiments were not sensitive to the context of the instructions and questions given to children and to their wording (Donaldson 1978).

In research on logical reasoning, significant differences were found between both children's and adults' responses, depending on the wording of questions, for example, in a task related to finding conditions in which a certain rule of the "p implies q" type is not satisfied (corresponding, in logic, to the negation of the implication). In one of the experiments, the subjects are shown four two-sided cards, which have a letter written on one side and a whole number on the other. Only one side of each card is visible, and the subjects see four symbols: A, G, 2, 5. The task is to verify in the fewest number of steps the rule "If there is a vowel on one side, there is an even number on the other." The rate of success of adults (usually university students) in this type of task is very small, 10 percent (Girotto 1989, p. 197). The success rate rises dramatically, to 90 percent, when the task is changed to deal with a less abstract context, for example, envelopes and stamps, and the rule to verify is "If the envelope is sealed, then the postage must cost 50 cents."

Contrary to Piaget's assumptions, adults are no more in tune with the rules of formal logic than children are. People do reason according to certain rules, but these rules are not the rules of the formal propositional calculus. They are not thinking in terms of premises and conclusions whose only meaning is in their truth value. In everyday life, people (children and adults) do not let their decisions be guided on the basis of abstract "if ... then" propositions that use the neutral verb *is;* they refer to emotion-laden rules containing such verbs as *must* or *mustn't* or *need.* These rules specify what is permitted to be done and what should not be done; children take such rules into account when they consider the consequences of something they plan to do (ibid.; Cheng and Holyoak 1985).

The apparent incompatibility between mathematical logic and the pragmatic logic of everyday language does not lead Navarra and Arzarello (chapters 19 and 15, respectively, in this volume) to the conclusion that the latter is some kind of obstacle that has to be overcome and replaced by the logic and syntax of mathematics. They claim that the development of expression in the natural language and the use of verbal code in problem solving can play important roles in the acquisition of more formal logical and algebraic modes of thinking. The relation between the ordinary language and mathematical thought is much more subtle than once believed.

Vygotsky seemed to be quite sensitive to the subtle interplay between natural language and spontaneous thought and scientific concepts. He was also one of the earliest critics of Piaget. The next section discusses his and his followers' views on language and communication.

THE SOCIOHISTORICAL APPROACH TO LANGUAGE AND COMMUNICATION: TEACHERS TALKING, STUDENTS LISTENING

Communication Is a Fact; Language Is Primarily an Instrument of Communication

For Vygotsky, communication is a cultural fact. If civilizations exist, it is thanks to the possibility of communication and handover (Bruner 1983; Edwards and Mercer 1993, pp. 23, 130) of the responsibility for knowledge and values from one generation to the next. Language is a cultural tool, a specifically human instrument of communication. In the course of an individual's development, language primarily follows this sequence: social speech, egocentric speech, internalized speech (or verbal thought). In the development of thought there is a preverbal stage, and in the development of speech there is a preintellectual stage. But when a child reaches about the age of two, his or her lines of thought and language start crossing each other. Thought becomes more verbal; speech becomes more intellectual. It is at this point that "[the child] seems to have discovered the symbolic function of words" (Vygotsky 1962, p. 43) and starts asking for names of things. From this point on, thought and language enhance each other's development.

This mutual relationship between thought and linguistic competence is assumed in Navarra's account of a study on children's logical reasoning (chapter 19 in this book). In this research, children were engaged in activities demanding "a constant interplay between the semantic and the logical aspects" of logical problems presented in the form of stories. It is assumed that by setting certain quite demanding standards on the forms in which students express themselves and justify their claims, it is possible to bring the children's logical thinking to a higher level.

Education and the Development of Thought Are Interdependent: Development as a Process of Enculturation

For Vygotsky, development does not mean just the growth and equilibration of abstract cognitive structures. It is related also to the contents of thought: what one thinks about and how one thinks about it. This view of development necessarily links it to education.

Intellectual development, in Vygotsky's sense, is the development of *concepts*. Concepts are understood as "meanings of words" (Vygotsky 1962, p. 112), but meaning is not viewed in a representational manner. Words do not depict or represent the experiential world. Whatever passes through the filter of speech loses its direct links with reality; "every word is already a theory" (Kozulin 1990, p. 88). This assumption underscores the discursive character of knowledge.

> [T]he French philosopher Henri Lefebvre ... pointed out the domination of symbols over referents in twentieth century everyday life. With the breakdown of traditional life in which referents were rather rigidly tied to symbols and concepts, and with the proliferation of mediated and the decline of first-hand experience, in the life of modern man symbols have assumed the role of ultimate reality. Communication and understanding depend more and more on a shared symbolic code, rather than on shared experience with the referents themselves. (Kozulin 1990, p. 164)

All this links thinking and development to language and communication, and thus to culture, in a way that is more causal than contingent.

The development of concepts in the child, from syncretic heaps, through complexes and pseudoconcepts, to fully fledged concepts at the age of adolescence, is indeed a process of enculturation: the learning of a discourse or even a range of discourses used in different spheres of practice, be it home, play, school subjects, or professions.

Vygotsky claims that mental functions depend on the nature of the things they operate on, or on what one thinks about, and hence on what one knows. The function of generalization depends on what is being generalized. Abstraction is a different process when applied to experiential reality than when it operates on some mental reality. This is why mental functions develop and change in the course of life and experience, and this is why their development is enhanced through schooling. Scientific questions, dealt with in formal education, give a totally different food for thought than do the problems encountered in everyday life.

Vygotsky's Criticism of Piaget

Piaget separated development and education, seeing the former as being not even partly derived from the latter. According to Vygotsky, the reason is that for Piaget, "mental functions do not change in the course of a child's development; only structures change and according to this the function acquires a new character" (Vygotsky 1989, p. 309).

Vygotsky sees an internal contradiction in the Piagetian system in relation to the separation of development and education.

His argument goes as follows. One of the basic principles of Piaget's theory is that the intellectual development of the child consists in a progressive socialization of thought that is, at the beginning, centered on itself, or "egocentric." But it is mainly at school that socialization of thought takes place. Piaget assumes, however, that school education can have no impact on development. This implies further that the non-spontaneous concepts of the child (mainly those that he or she systematically learns at school) do not reflect the specific features of his or her thinking, which are present only in the spontaneous concepts. Thus, the spontaneous and the nonspontaneous concepts are separated by a barrier that excludes the possibility of a mutual interaction between these concepts (Vygotsky 1962, p. 85).

> [According to Piaget] throughout childhood, there is a ceaseless conflict between the two mutually antagonistic forms of thinking [the spontaneous and the non-spontaneous], with a series of compromises at each successive developmental level, until adult thought wins out. The child's own nature plays no constructive part in his intellectual progress. When Piaget says that nothing is more important for effective teaching than a thorough knowledge of the spontaneous child thought ... he is apparently prompted by the idea that child thought must be known as an enemy must be known in order to be fought successfully. (Vygotsky 1962, p. 85)

The claim, mentioned above by Vygotsky, that for Piaget, "the intellectual development of the child consists in a progressive socialization of his thought" is not in contradiction with the implication that communication is an impossible concept in Piaget's theory. Socialized thought is no more communicable than nonsocialized thought because the socialization of thought is a process that goes from inside the developing subject. The thought is socialized when the subject starts thinking in terms of how he or she is going to communicate his or her thoughts to others so that these others understand and accept his or her point. This is not the same as saying, for example, that socialization of thought occurs because the person has participated in a culture and now starts sharing certain ways of thinking with other members of the culture.

The Role of Writing in the Development of Mental Functions

In stressing the role of education for development, Vygotsky devoted much attention to the acquisition of written language and its merits for the development of thought (Vygotsky 1962, p. 98ff.). The arguments he was using were very much the same as those to which Piaget resorted as he compared egocentric thought with socialized thought, or thought for the purposes of communication to others. However, Vygotsky was claiming that writing can have an actual impact on development; Piaget would not say that the activity of communication can change the course of development. On the contrary, he would have claimed that development is a precondition for a person to express himself or herself clearly in writing.

These subtleties of theoretical assumptions do not make much difference for the practice of education, and we see both constructivists and Vygotskian researchers promote writing in mathematics classrooms. What is understood under this phrase today, however, may not have always crossed Vygotsky's mind: writing diary entries in mathematics classes or describing one's heuristic methods in solving a mathematical problem. These are more recent inventions.

The most important feature of written speech for Vygotsky is its voluntary character, the fact that it is planned and conscious and is based on an arbitrarily chosen system of signs. The adoption of an alphabet is a conscious process, whereas the changes in spoken language (across time) remain unconscious in societies (Vygotsky 1989, p. 238). Also deliberate and conscious is the adoption of technical notations, such as the mathematical ones, which are often subject to debates (e.g., the debate between the Newtonian and the Leibnizian notation in calculus).

Thus, a totally different kind of thinking is involved in writing, and this has an influence on development. But writing is not something that a child would learn without someone consciously and systematically teaching him or her. The spoken language can be learned and taught in an unplanned and not very systematic manner; writing requires special training.

An Aim of Education: To Develop a Conscious Reflection on Cultural Practices, Such as Speech and the Use of Numbers, That in Everyday Life Are Often Taken for Granted

The child is learning the written language not to speak better but to think better. The child does not have to have already developed better thinking to be able to express himself or herself in writing. The writing and the development of thinking will mutually enhance each other. Likewise, the child is not learning grammar to speak better:

> A kindergarten child has already mastered the basic forms of grammar and syntax. During the lessons of mother tongue at school, the child is not acquiring new grammatical habits or syntax structures. From this point of view the learning of grammar would be indeed useless. What the child is learning at school through writing and grammar is to become conscious of what he is doing; he is learning to use his abilities in a voluntary way. The ability [to use language] is transferred from the level of unconscious and automatic action to the level of a conscious and planned action. (Vygotsky [1989, p. 242], translated from Polish)

The aim of learning mathematics at school is similar. It is not to become better at doing sums or divisions but rather to understand how they are done. The child is supposed to learn to use these operations in a conscious and voluntary way.

> The generalization of one's own arithmetic operation is something higher and new compared with the generalization of the quantitative features of objects.... The lower operation starts to be considered a particular case of the higher operation. (Vygotsky [1989, p. 294], translated from Polish)

This new generalization leads to algebraic thinking in the light of which arithmetic relationships between numbers are particular cases of general laws represented in terms of variables.

A Pedagogy Grounded in Historically Relevant Social Practices

Vygotsky's ideas about the role of schooling and the place of written language in the development of mental functions were taken up by followers of Vygotsky, especially Davydov (1982) (cited in Cobb, Perlwitz, and Underwood [1994]; Lompscher 1994), whose teaching designs embody certain Vygotskian assumptions about the historical and social character of human experience and the importance of labor. Achievements of human intellect are derived from various social practices and are transmitted from generation to generation in a symbolically mediated way, not through biological transmission (Vygotsky 1925–1979; for a comprehensive comparison between the constructivist and Vygotskian positions, see Confrey [1994, 1995]). Relevant social practices give meaning to, and justify the existence of, particular domains of scientific research and their theoretical frameworks.

For example, the notion of real number as a ratio can be derived from the practice of measuring and comparing sizes of objects, which can be understood to include counting when the compared objects are sets of other objects. Such was the basis of several pedagogical experiments conducted by such researchers as Galperin, Talysina, and Davydov (Lompscher 1994). In particular, the design of Davydov's large-scale and long-term experiments in the 1960s was targeted toward the first three primary grades. There were several steps in the introduction of children to the world of numbers. The first two, as described by Lompscher, were as follows. First, the concepts of equal, larger, and smaller (and their symbols: =, >, <) were introduced in the context of situations involving a direct comparison of objects with respect to such features as length, breadth, height, weight, and area by juxtaposition, superposition, and so on. In the second step, the objects presented to students were too difficult to be compared directly, which justified the use of measuring sticks, threads, and other devices. Students were introduced to using symbolic representations of relationships, such as $A/c = 5$ (the measure c is contained 5 times in the quantity A).

This discussion already gives an idea of the approach and its heavy reliance on the introduction of culturally accepted, conventional mathematical symbolism. Piaget was very much against such early imposition of conventional mathematical symbolism on children; he preferred that children be allowed to use representations of their own invention, even if understood only by themselves. Let us mention, however, that in the foregoing so-called sociohistorical approach to teaching number, the symbolism is not just imposed on children. Care is taken for this symbolism to be linked with a widely shared cultural practice, in this instance the practice of measuring and comparing quantities, about which children are not only informed but in which they are actually and actively engaged. The symbolism thus appears as a useful and convenient tool in keeping a record of the results of measurement and, when further developed, in deducing facts or predicting relations not available through direct measurements. Here, the Vygotskian approach meets the constructivist notion of knowledge, as an answer to a problem that is regarded as one's own problem.

In designing the teaching of a mathematical concept, a constructivist looks for problem situations for whose solution the concept would be an optimal tool, whereas a proponent of Vygotsky's theory tries to locate cultural practices that, historically, gave rise to the concept in question. Since the chain of links that relate abstract mathematical concepts with such cultural practices is usually very long, the Vygotskian approaches tend to be based on long-term teaching scenarios, starting in early grades and extending to secondary and higher education. They are also described in terms of activities proposed to students, and teacher interventions are *not*, a priori, specified as having to *depend* on the students' possible reactions and spontaneous understandings. The teacher is not expected to follow the student's lead as in constructivist approaches (Arcavi and Schoenfeld 1992); rather, the students are expected to follow the teacher's lead. Bartolini Bussi, in chapter 3 in this volume, refers to this principle as "meanings of scientific concepts are not to be negotiated" and discusses the pedagogical difficulty it creates in view of another claim of the theory that "meanings cannot be taught directly."

Classroom Communication Styles: The Question of the Transmission of Knowledge

Vygotsky distinguished between the spontaneous concepts that the child develops through informal interactions with people in everyday life and the scientific concepts that the child learns at school. The spontaneous concepts may be inductive generalizations from the everyday experiences of the child, but the development of scientific concepts goes in the opposite direction: from the general to the concrete (Kozulin 1990, p. 169).

Thus in the Vygotskian vision of the mathematics classroom, it is normal to have the students study a concept's definition as a starting point for the acquisition of this concept; they would be expected to analyze its logical structure; to find examples and nonexamples of the concept; and to figure out its place in the structure of a theory—its possible applications and so on. In a constructivist classroom, the teacher would start by giving students a problem for which the concept in question appears to be an optimal tool to use in its solution. The concept would have to be constructed from within the individual student's mental structures as an answer to a problem that he or she considers his or her own. The study of existing texts would come later on or not at all. Metaphorically speaking, in a constructivist classroom, each student writes his or her own text of knowledge.

But Vygotsky believed in the possibility of the verbal transmission of knowledge no more than did Piaget. First of all, the mental functions (such as voluntary memory, logical memory, abstraction, comparison, and discrimination) cannot be learned as recipes. They cannot be learned in a ready-made form (Vygotsky 1989, pp. 169–70). Also, the meanings of words cannot be transmitted from one mind to another with the help of other words because every word is a generalization and different people may operate with different levels of generalization as a mental function (ibid.). However, Vygotsky rejects the claim that "it is impossible to intervene into this mysterious process [of concept formation] and that the development of concepts should be left to its own internal laws" (Vygotsky 1962, p. 83). The intervention must only be much more subtle and more indirect than the vulgar methods of teaching by rote memorization of rules and formulas. This does not mean that teachers should refrain from giving students definitions or verbal explications of terms. It only should not be expected that when, for example, a child is told what a word means, he or she absorbs the definition or explication and from this point on has the concept. This must be regarded as the beginning of the process of the development of the concept and requires serious work and reflection on the part of the child (Vygotsky 1962, p. 108).

A Transition to Interactionism

Both Piaget's and Vygotsky's theories concentrate on progress: The human intellect is seen as developing from the child's immature thought and language to the adult's decontextualized logic and scientific concepts. Children's speech and thought are analyzed from the point of view of their shortcomings with respect to adult or scientific speech and thought.

Mathematics educators claiming links with these two psychologists often also speak of what classroom communication should look like and speak of the children's actual linguistic behaviors and classroom communication in terms of how they could be corrected. Few constructivists have tried to explain why the classroom communication is as it is.

This question was tackled by other groups of researchers. The outcome of their investigations and reflection is not so much a new pedagogy for the teaching of mathematics but conclusions of the following type: "if teachers and students engage in interactions of type A, then students are likely to develop ways of knowing and understanding of type $f(A)$." Teachers who are not happy with having their students develop this type of understanding and knowing may have to change the way in which they interact with their students (cf. Steinbring [1993] and chapter 5 in this volume). These researchers tend to draw on Bruner's theory of language acquisition (Bauersfeld 1994; Krummheuer

1995; Voigt 1995), on pragmatic approaches to language and meaning (Pimm 1994), and on Wittgenstein's later work (Gergen 1995).

In particular, two of Wittgenstein's ideas are often evoked. One is "the meaning of a word is in its use" (Wittgenstein 1969, pp. 4, 65). It is discussed in depth in Kanes's chapter in this book. The other is "ordinary language is all right" (ibid., p. 28); one should try to understand it rather than show its logical fallacies and call for it to be corrected. This remark inspires a more standoffish attitude toward the reality of mathematics classrooms, the attitude of an informed and impartial observer and not one of a missionary or a social worker who tries to "organize people's lives for them" (Robert Park, cited in Hammersley [1989, p. 79]). Of course, change is envisaged as an ultimate goal, but it is believed that this change must be prepared on the grounds of a profound understanding of the persistence and reproducibility of the existing culture (see Steinbring, chapter 5 in this volume).

In the mathematics education literature, approaches that display some kinship with the foregoing ideas identify themselves as "social constructivism," "social constructionism," "interactionism," and "alternative epistemologies" (Steffe and Gale 1995). Disregarding certain subtle theoretical differences among these approaches, we shall label them all "interactionism" and have a look at how interactionism views language and communication.

INTERACTIONIST APPROACHES TO LANGUAGE AND COMMUNICATION: TEACHERS AND STUDENTS IN A DIALOGUE

Communication and Language in the Foreground of Research

As Gergen (1995) rightly points out, for all Vygotsky's stress on the social factor in human development, communication itself remains in the background of his and his followers' research. Like Piaget, Vygotsky studied the development and the workings of the single human mind—such has been, in fact, the preoccupation of traditional psychology and genetic epistemology. There exists, however, another current of psychological research for which interactions, and not the psychological subject, have become the object of study. Certain types of interactions have received special attention, for example, the interactions between the genome and the environment—or the interactions between genetic endowment and experience in the course of development (including the famous nature-nurture problem)—and the interactions between the individual and the culture. The latter include interactions between a developing child and an adult; interactions between a learner, or a group of learners, and a teacher; as well as interactions between peers (Forgas 1985; Bruner and Bornstein 1989a, 1989b). Research has also been devoted to the relations between language acquisition and the acquisition of knowledge about the social world and the world generally (Bruner 1985).

In interactionist approaches focusing on individual-culture interactions, language has an important place. This emphasis is related to the notion of knowledge that these approaches normally hold. The underlying epistemology locates the sources of the validity of knowledge not in the observation of the objective world (empiricism) or in innate rationality (rationalism) or even in the logico-mathematical structures of the mind constructed through a sequence of developmental stages (constructivism), but in language, where language is understood not as a system of signs but as a social practice—a discourse (Sierpinska 1994, pp. 17–21). It is language-in-action, or language as a means for accomplishing cognitive, social, and other ends. It is a "vehicle for doing things with and to others" (Bruner 1985). Thus knowledge is seen as having a discursive character.

Mathematics is therefore also a discourse, and, as such, it is not just a tool for solving problems, but something much more influential. It is a way of seeing the world and thinking about it. It is a universe that is established through communication, whereby people commit themselves to certain *conventions*, build shared understandings of *contexts*, and develop "conventional means for jointly establishing and retrieving *presuppositions*" (Bruner 1985, p. 38). Thus, mathematics is a language seen from the perspective of pragmatics, not semantics or syntax.

The fruitfulness of the linguistic-pragmatic perspective for research in mathematics education has been advocated by several authors (e.g., Pimm [1994]; Girotto[1989]).

Language as a Social Practice

Language, from the interactionist perspective, is primarily an instrument of communication, but not in the sense of communication of thoughts. For the interactionist, the term *communication* can be used in sentences like these: "Two people communicate with each other" or "This group of students communicates about a problem." Within a constructivist frame of mind or in sociocultural approaches, one would be expected to use this verb in a transitive form: "A person A communicates K to a person B," where K can be a thought or a piece of historically and culturally established knowledge.

Interactionists are not making claims about the correspondence or lack thereof between what people say and what they think. People are "doing things with words" (Austin 1975); in speaking, they are "doing things with and to others" (Bruner 1985, p. 38), and it is through this shared activity that meanings are spread in a culture. Meanings are not in people's heads, to be transmitted from one person to another. People do not have to mean what they say, but what they say definitely means something, not just "to others" but something in the given culture.

If this view of meaning is to be seriously taken into account, then language cannot be spoken about as something independent of its use in communication. It is not a stand-alone tool for which one reaches when one wants to express a thought. One does not really speak a language, one is "languaging" in a language (Bauersfeld 1995, p. 272). "Speaking a language" conveys the idea that it is possible to say exactly the same thing by using two different languages, the language being just a medium through which it would be possible to express one's thoughts. But our thoughts can change with the language we happen to be speaking at any one moment (ethnic, technical, ordinary, quasi-mathematical), and "languaging" is not just, or not at all, "communicating one's thoughts"—it is "a way of life" (Wittgenstein [1958], cited in Bauersfeld [1995, p. 279]).

The interactionist view of communication is associated with the school of philosophy of language represented by later Wittgenstein, Austin, Searle, Grice, and their followers. In this approach, "meaning" is related to use and intention, as opposed to the approach, represented by Frege and early Wittgenstein, where it is associated with truth conditions (Searle 1986). Gergen expresses well these assumptions of interactionism about language when he states that "meaning in language is achieved through social interdependence" and that "meaning in language is context dependent," as well as that "language primarily serves communal functions" (Gergen 1995, pp. 24–26).

The impact that Wittgenstein's view of language can have on the way we interpret mathematics classroom interaction is demonstrated and discussed in chapter 6 by Kanes.

Meaning Is in the Discourse

Interactionism thus endorses a notion of meaning that locates it in the language seen as a discourse. Meanings are generated through discourse in the regulation of practices. Interactionism rejects the "representationist" view of meaning according to which the relationship between the signifier and the signified is one of representation and the signifier is understood as representing the signified in the subject's mind. This view is believed to have been held by Piaget. For Piaget, says Walkerdine (1988, pp. 3–4), "the signified is something extra-linguistic, something arising extra-discursively; the schemata represented by mathematical signifiers, e.g., numerals, have their origin not in the subject's relation to a system of signs as a social phenomenon, but ultimately in the coordination of actions whose function is successively to equilibrate the biological subject and its environment."

Walkerdine believes in the primacy of the signifier over the signified—"we live in a world of words, where reality always eludes us." This is an "ideological" view of meaning, to which she adds a historical dimension, adopted from Foucault's poststructuralist position toward language (Foucault 1978): The world is not known except through discourse and discourse is based on sign systems that are not universal and transhistoric but specific, historically generated bodies of knowledge.

This ideological view of meaning does not see ideology in an evaluative and pejorative way. Ideology is not understood as a politically enforced system of unjustified claims that distort reality, as opposed to "truthful science." Ideology is a world view that is there, whether we want it or not; it is an aspect of the discourse (cf. Noss [1994]). Walkerdine's message is *not* that one discursive practice is distorting the meanings of words and expressions that are also used in another practice. For example, at home, the child hears and uses the word *heavy* most often in reference to objects that the child cannot carry, for example, in the expression *too heavy:* "it is too heavy for you to carry." It thus refers to the child's strength. At school this same word is used as an antonym of *light* and is used in speaking about quantitative qualities of material objects. Home use does not distort the true meaning of the word; it is just used differently there because home represents a specific sphere of relations between adults and children, different from that established at school. Topics of conversation and goals of interactions are different between home and school. Discrimination between heavy and light things at home is usually only a means to achieve some practical purpose. At school this discrimination becomes the topic of a lesson, and children are explicitly discussing the meanings of the words *heavy* and *light*. To survive at school, the child must notice the differences between discourses and put his or her efforts into becoming a practitioner of the new discourse.

The Fundamental Role of Communication in Development and Education

In psychogenesis, interactionism sees communication as preceding and preparing the ground for language acquisition. In this we refer to Bruner's theory:

> [L]anguage acquisition "begins" before the child utters his first lexico-grammatical speech. It begins when mother and infant create a predictable format of interaction that can serve as a microcosm for communicating and for constituting a shared reality. The transactions that occur in such formats constitute the "input" from which the child then masters grammar, how to refer and mean, and how to realize his intentions communicatively. (Bruner 1985, p. 31)

Before the child utters a phrase like "Bye, bye, come back soon," he or she has already participated in situations of a familiar person's leaving and he or she expects certain gestures to be made and a certain kind of verbal exchange to take place. The child has been acquainted with the *format* of farewell.

Bruner's important concept of interaction format evoked above has been extended and adapted to mathematics education by Krummheuer in chapter 13.

The Type of Knowing Depends on the Type of Classroom Interaction

If what the student learns as mathematics is a certain discourse, then his or her way of knowing mathematics is a function of the characteristics of the communication and interactions in which he or she participates in the process of learning. These characteristics include the following factors. Who sets the agenda for class activity, who decides what is relevant in a task, who provides the concepts—the teacher or the students? What is the subject of communication—a computational procedure, an interpretation of a concept in terms of possible interpretations, the formal structure of a definition, the modeling of a real-life situation in mathematical terms, a technique of solving a range of typical test problems? What interaction format establishes itself and evolves among the actors in the communication, what is the dominant routine or the predictable scheme or scenario of interaction—is it interrogation or is it an interview, is it the training of skills or is it the elicitation of reflection, is it repetition and recitation or mutual challenging through provocative questions? In each situation, a different discourse will be understood and learned. The issue of what is learned in different *formats* of classroom interaction is addressed especially by Wood and Krummheuer, chapters 9 and 13, respectively.

Communication in the Domain of Mathematics

Communication in the domain of mathematics is difficult for those who participate in the process not because the transmission of thoughts from one mind to another is fallible or impossible, but because the specificity of mathematics itself imposes stringent demands on communication.

Mathematical communication is bound to the use of linguistic means—there is absolutely nothing in mathematics that can be *shown* (although mathematical texts are full of this verb). As has often been repeated, mathematics is about relations, not about things, and relations cannot be experienced directly.

Steinbring (1994, p. 94) draws our attention to what he calls "an epistemological dilemma in every mediation of mathematical knowledge." In trying to teach a new mathematical topic, the teacher introduces a new language: new terms, symbols, definitions. He or she stresses the conventional and technical rules of the use of this new language. The students may thus be drawn to focus on the concrete aspects of the notation, although the mathematical ideas behind the notation may elude them.

The meanings of mathematical expressions—words, formulas, diagrams—are found only when they become part of a discourse that the student shares with others. They are found when the student starts to use the new language and realizes that through its use, he or she can actually do something to and with others and achieve certain goals about which he or she cares. The meaning of a symbol is in the shared and actually used discourse as a whole. It is the discourse as a whole that lends meaning to its parts, not the other way around.

One special feature of the mathematical written language is its two-dimensionality. As Peirce noted, "any algebraic expression is essentially a diagram" (cited in Otte [1983, p. 17]). This is, in particular, why mathematics cannot be taught by telling—speech is unidimensional. (This argument does not hold if telling is accompanied by gestures suggestive of spatial metaphors and by modulation of voice; in that situation, we have a multidimensional language.)

Mathematical texts are not read in a linear fashion. There are constant returns in search of hidden clues. A formula has to be read from left to right and from right to left and grasped as an almost iconic whole. Expressions with quantifiers require the reader to make multiple returns to the quantifiers, to check for free variables and to become aware that their order is essential.

Mathematical texts are fraught with diagrams and graphs that are two-dimensional texts in a more explicit fashion. For these to become part of a shared discourse between teachers and students, a lot of focusing on the part of the teacher, and work and experience on the part of the students, is needed. Graphical representations invariably contribute to the difficulty of learning mathematics, while they constitute an irreplaceable prop for its understanding. Otte (1983) ascribes to Peirce the view that "the only way of communicating an idea [is] to do so by the visual means of icons or diagrams." They are misleading through their pictorial character, but their conciseness, once grasped, allows invaluable shortcuts for thinking and solving problems. Chapter 10 by Curcio and Artzt and chapter 17 by Kaldrimidou and Ikonomou discuss aspects of this problem.

An additional but very common difficulty in communication arises when, in trying to facilitate the participation of students in the mathematical discourse, the teacher stresses the conventions of proper writing. For example, some teachers stress a certain type of presentation for the solution of an equation or a system of equations. Each step must be written in a new paragraph, the same variables in a system should be written exactly one under the other, all operations on the equations (adding something to both sides of the equation or multiplying, etc.) must be explicitly mentioned on the side, and so on. All these rules are aimed at helping the student avoid mistakes by making the solution more orderly; by keeping track of all operations done, the student is able to detect a mistake if he or she has made one. But students do not know, from the outset, what is relevant to the concept of equation and what is relevant for the practice of finding solutions, and everything becomes relevant on the same level (Skovsmose, personal communication; see also Alvø and Skovsmose [1996]). Of course, at school, it is important to find the right solution to an equation because this skill is often the basis on which the overall mathematical ability of the student is evaluated. Thus the technique of solving equations efficiently is important, and teachers devote much of their creative thinking to this question.

To improve communication, textbook authors and teachers resort to a variety of means, among which metaphors and analogies are perhaps the most common. A well-known one is the metaphor of balance, and especially the metaphor of scales in the teaching of solving equations (Otte 1983, pp. 20–21). This metaphor, however, does not help the students see equations in a broader conceptual perspective. It only, maybe, helps them get some understanding of the technique. But it does not help them understand the use of equations in thinking mathematically about relaticnships. It does not help them view equations as a specific class of conditions on variables among other possible types of conditions. But the language of scales in the context of equations has become part of school mathematics. Students have to learn this language, and this creates problems of its own, as clearly shown by Pirie (chapter 1) and MacGregor (chapter 16). This fact is an example of a broader phenomenon, called "glissement métacognitif" by Brousseau (1986, p. 291). A linguistic device used to help students understand a mathematical topic becomes an object of teaching in itself; the classic example is that of Venn diagrams used in the context of sets in elementary schools during the New Math reform period.

An "Interactionist Pedagogy"

From the interactionist perspective, transmission of knowledge is not an issue because knowledge is not in the head of the teacher. It is something that emerges from shared discursive practices that develop within the cultures of the classroom, the school institution, and the society at large. Interactionist epistemological assumptions per se do not imply pedagogical principles, but interactionists entertain visions of classroom cultures in which more valuable mathematics could be learned. They do not always share the same understanding of what counts as valuable mathematics.

A social constructionist view, represented by Gergen, paints a broad vision of what general education should look like and what would be the place of mathematics in it.

[L]anguage acquires both its social value and its meaning largely from the way in which it is used by people in specific contexts. The challenge for the educational process, then, is not that of storing facts, theories, and rational heuristics in individual minds. Rather ... it is to generate the kinds of contexts in which the value and meaning of the constituent dialogues may be most fully realized, conditions under which dialogues may be linked to the ongoing practical pursuits of persons, communities, or nations.... [T]he constructionist would favor practices in which students work together with teachers to decide on practical issues that are important to them.... For example, if students are concerned about ecology, racial tension, abortion, drugs, and so on, can they develop projects that will elucidate the issues, and can they communicate their insights and opinions effectively to others? ... For the constructionist, educational dialogues should be wedded as closely as possible to the circumstances of application. To put it in other terms, why should education be preparatory to communal existence rather than a significant form of existence.... [M]athematical techniques may become the needed tools of understanding and expression—for determining the significant rise and fall in various phenomena, for assessing costs and benefits, for reading demographic charts, or for effectively communicating the results of one's studies to others. (Gergen 1995, pp. 35–36)

This vision of teaching mathematics in contexts is not something that any interactionist mathematics educator would endorse without reservations (Sierpinska 1995), although everybody is concerned with the question of transfer of knowledge. It remains to be seen, and systematic research needs to be done on what kinds of meanings and ways of knowing are developed in classrooms in which mathematics is taught "in contexts," "through applications," or "through examples and problem solving." Would we be satisfied with this type of knowledge?

The interactionist approach opens up avenues for research that are fascinating for mathematics educators because they take so much of the experientially known reality of the mathematics classroom into account. It focuses on what for us as teachers is the main occupation and the biggest challenge: our interactions with students in and about mathematics.

ACKNOWLEDGMENTS

The writing of this chapter was supported by the Canadian Social Sciences and Humanities Research Council (grant no. 410-93-007). All opinions expressed are solely those of the author.

REFERENCES

Alvø, H., and O. Skovsmose. "Students' Good Reasons." *For the Learning of Mathematics* 16 (1996): 31–38.

Arcavi, A., and A. H. Schoenfeld. "Mathematics Tutoring through a Constructivist Lens: The Challenges of Sense-making." *The Journal of Mathematical Behavior* 11 (1992): 321–36.

Austin, J. L. *How to Do Things with Words.* Cambridge: Harvard University Press, 1975.

Bauersfeld, H. "Theoretical Perspectives on Interaction in the Mathematics Classroom." In *Didactics of Mathematics as a Scientific Discipline,* edited by R. Biehler, R. Scholz, R. W. Sträßer, and B. Winkelmann, pp. 133–46. Dordrecht, Netherlands: Kluwer Academic Publishers, 1994.

———. "'Language Games' in the Mathematics Classroom: Their Function and Their Effects." In *The Emergence of Mathematical Meaning: Interaction in Classroom Cultures,* edited by P. Cobb and H. Bauersfeld, pp. 271–91. Hillsdale, N. J.: Lawrence Erlbaum Associates, 1995.

Bergeron, J., N. Herscovics, and N. Nantais. "Formative Evaluation from a Constructivist Perspective." In *Proceedings of the Seventh Annual Meeting of the North American Chapter of the International Group for the Psychology of Mathematics Education,* pp. 13–19. Columbus, Ohio: International Group for the Psychology of Mathematics Education, 1985.

Brousseau, G. "Fondements et méthodes de la didactique des mathématiques." *Recherches en didactique des mathématiques* 7 (1986): 33–115.

Bruner, J. S. *Child's Talk: Learning to Use Language.* Oxford: Oxford University Press, 1983.

———. "The Role of Interaction Formats in Language Acquisition." In *Language and Social Situations,* edited by J. P. Forgas, pp. 31–46. New York: Springer-Verlag, 1985.

Bruner, J. S., and M. H. Bornstein. "On Interaction." In *Interaction in Human Development,* edited by J. S. Bruner and M. H. Bornstein, pp. 1–14. Hillsdale, N.J.: Lawrence Erlbaum Associates, 1989a.

Bruner, J. S., and M. H. Bornstein, eds. *Interaction in Human Development.* Hillsdale, N.J.: Lawrence Erlbaum Associates, 1989b.

Cheng, P. W., and K. J. Holyoak. "Pragmatic Reasoning Schemas." *Cognitive Psychology* 17 (1985): 391–416.

Cobb, P. "Where Is the Mind? Constructivist and Sociocultural Perspectives on Mathematical Development." *Educational Researcher* 23 (1994): 13–20.

Cobb, P., M. Perlwitz, and D. Underwood. "Construction individuelle, acculturation mathématique et communauté scolaire." *Constructivisme et éducation: Revue des sciences de l'éducation.* Numéro thématique 20 (1994): 41–61. Special issue edited by M. Larochelle and N. Bednarz.

Confrey, J. "What Constructivism Implies for Teaching." In *Constructivist Views of the Teaching and Learning of Mathematics*, edited by R. B. Davis, C. A. Maher, and N. Noddings, pp. 107–22. *Journal for Research in Mathematics Education* Monograph Series, no. 4. Reston, Va.: National Council of Teachers of Mathematics, 1990.

———. "A Theory of Intellectual Development (Part I)." *For the Learning of Mathematics* 14, no. 3 (1994): 2–8.

———. "A Theory of Intellectual Development (Part II)." *For the Learning of Mathematics* 15, no. 1 (1995): 38–47.

———. "A Theory of Intellectual Development (Part III)." *For the Learning of Mathematics* 15, no. 2 (1995): 36–47.

Davis, R. B., C. A. Maher, and N. Noddings, eds. *Constructivist Views of the Teaching and Learning of Mathematics*. *Journal for Research in Mathematics Education* Monograph Series, no. 4. Reston, Va.: National Council of Teachers of Mathematics, 1990.

Davydov, V. V. "The Psychological Characteristics of the Formation of Elementary Operations in Children." In *Addition and Subtraction: A Cognitive Perspective*, edited by T. P. Carpenter, J. M. Moser, and T. A. Romberg, pp. 224–38. Hillsdale, N.J.: Lawrence Erlbaum Associates, 1982.

Donaldson, M. *Children's Minds*. London: Fontana Books, 1978.

Edwards, D., and N. Mercer. *Common Knowledge: The Development of Understanding in the Classroom*. London: Routledge, 1993.

Forgas, J. P., ed. *Language and Social Situations*. New York: Springer-Verlag, 1985.

Foucault, M. *The Order of Things. An Archeology of Human Sciences*. New York: Vintage Books, 1973.

Gergen, K. J. "Social Constructionism and the Educational Process." In *Constructivism in Education*, edited by L. P. Steffe and J. Gale, pp. 17–40. Hillsdale, N.J: Lawrence Erlbaum Associates, 1995.

Girotto, V. "Logique mentale, obstacles dans le raisonnement naturel et schémas pragmatiques." In *Construction des savoirs: Obstacles et conflits*, edited by N. Bednarz and C. Garnier, pp. 195–205. Ottawa: Agence d'Arc Editions, 1989.

Hammersley, M. *The Dilemma of Qualitative Method. Herbert Blumer and the Chicago Tradition*. London: Routledge, 1989.

Kilpatrick, J. "What Constructivism Might Be in Mathematics Education." In *Proceedings of the Eleventh Conference of the International Group for the Psychology of Mathematics Education*, edited by J. C. Bergeron, N. Herscovics, and C. Kieran, pp. 2–27. Montreal: International Group for the Psychology of Mathematics Education, 1987.

Kozulin, A. *Vygotsky's Psychology: A Biography of Ideas*. New York: Harvester-Wheatsheaf, 1990.

Krummheuer, G. "The Ethnography of Argumentation in the Mathematics Classroom." In *The Emergence of Mathematical Meaning: Interaction in Classroom Cultures*, edited by P. Cobb and H. Bauersfeld, pp. 229–69. Hillsdale, N.J.: Lawrence Erlbaum Associates, 1995.

Laborde, C. "Working in Small Groups: A Learning Situation?" In *Didactics of Mathematics as a Scientific Discipline*, edited by R. Biehler, R. Scholtz, R. W. Sträßer, and B. Winkelmann, pp. 147–58. Dordrecht, Netherlands: Kluwer Academic Publishers, 1994.

Larochelle, M., and N. Bednarz, eds. "Constructivisme et éducation." *Revue des sciences de l'éducation*. Numéro thématique 20 (1994).

Lerman, S. "Intersubjectivity in Mathematics Learning: A Challenge to the Radical Constructivist Paradigm?" *Journal for Research in Mathematics Education* 27 (1996): 133–50.

Lesh, R., and A. E. Kelly. "Action-Theoretic and Phenomenological Approaches to Research in Mathematics Education: Studies of Continually Developing Experts." In *Didactics of Mathematics as a Scientific Discipline*, edited by R. Biehler, R. Scholtz, R. W. Sträßer, and B. Winkelmann, pp. 277–86. Dordrecht, Netherlands: Kluwer Academic Publishers, 1994.

Lompscher, J. "The Sociohistorical School and the Acquisition of Mathematics." In *Didactics of Mathematics as a Scientific Discipline*, edited by R. Biehler, R. Scholtz, R. W. Sträßer, and B. Winkelmann, pp. 263–76. Dordrecht, Netherlands: Kluwer Academic Publishers, 1994.

Noss, R. "Structure and Ideology in the Mathematics Classroom." *For the Learning of Mathematics* 14 (1994): 2–10.

Otte, M. "Textual Strategies." *For the Learning of Mathematics* 3 (1983): 15–28.

Piaget, J. *The Language and Thought of the Child*. London: Routledge & Kegan Paul, 1959.

———. *Science of Education and the Psychology of the Child*. New York: Viking Press, 1972.

———. *Judgment and Reasoning in the Child*. Totowa, N.J.: Littlefield, Adams & Co., 1976.

Pimm, D. "Mathematics Classroom Language: Form, Function and Force." In *Didactics of Mathematics as a Scientific Discipline*, edited by R. Biehler, R. Scholtz, R. W. Sträßer, and B. Winkelmann, pp. 159–70. Dordrecht, Netherlands: Kluwer Academic Publishers, 1994.

Richards, J. "Construct[ion/iv]ism: Pick One of the Above." In *Constructivism in Education*, edited by L. P. Steffe and J. Gale, pp. 57–63. Hillsdale, N.J.: Lawrence Erlbaum Associates, 1995.

Searle, J. "Meaning, Communication and Representation." In *Philosophical Grounds of Rationality: Intentions, Categories, Ends*, edited by R. E. Grandy and R. Warner, pp. 209–26. Oxford: Clarendon Press, 1986.

Sierpinska, A. *Understanding in Mathematics*. London: Falmer Press, 1994.

————. "Mathematics: 'In Context,' 'Pure' or 'With Applications': A Contribution to the Question of Transfer in the Learning of Mathematics." *For the Learning of Mathematics* 15 (1995): 2–15.

Steffe, L. P., and J. Gale, eds. *Constructivism in Education.* Hillsdale, N.J.: Lawrence Erlbaum Associates, 1995.

Steinbring, H. "Problems in the Development of Mathematical Knowledge in the Classroom: The Case of a Calculus Lesson." *For the Learning of Mathematics* 13 (1993): 37–50.

————. "Dialogue between Theory and Practice in Mathematics Education." In *Didactics of Mathematics as a Scientific Discipline*, edited by R. Biehler, R. Scholtz, R. W. Sträßer, and B. Winkelmann, pp. 89–102. Dordrecht, Netherlands: Kluwer Academic Publishers, 1994.

Voigt, J. "Thematic Patterns of Interaction and Sociomathematical Norms." In *The Emergence of Mathematical Meaning: Interaction in Classroom Cultures*, edited by P. Cobb and H. Bauersfeld, pp. 163–202. Hillsdale, N.J.: Lawrence Erlbaum Associates, 1995.

von Glasersfeld, E. "The Reluctance to Change a Way of Thinking." *Irish Journal of Psychology* 9 (1988): 83–90.

————. "Pourquoi le constructivisme doit-il être radical?" *Revue des sciences de l'éducation* 20 (1994): 21–28.

————. "A Constructivist Approach to Teaching." In *Constructivism in Education*, edited by L. P. Steffe and J. Gale, pp. 3–16. Hillsdale, N.J.: Lawrence Erlbaum Associates, 1995.

Vygotsky, L. S. "Consciousness as a Problem of the Psychology of Behaviour." *Soviet Psychology* 17 (1979): 5–35. First published in 1925.

————. *Thought and Language.* Cambridge: M.I.T. Press, 1962.

————. *Myslenie i mowa.* Warsaw: Panstwowe Wydawnictwo Naukowe, 1989.

Walkerdine, V. *The Mastery of Reason: Cognitive Development and the Production of Rationality.* London: Routledge, 1988.

Wittgenstein, L. *The Blue and Brown Books.* Oxford: Basil Blackwell, 1969.

Wood, T. "From Alternative Epistemologies to Practice in Education: Rethinking What It Means to Teach and Learn." In *Constructivism in Education*, edited by L. P. Steffe and J. Gale, pp. 331–40. Hillsdale, N.J.: Lawrence Erlbaum Associates, 1995.

Part 2

Different Approaches to the Study of Communication in the Mathematics Classroom

3

Verbal Interaction in the Mathematics Classroom: A Vygotskian Analysis

Maria G. Bartolini Bussi

University of Modena

I N RECENT years several research papers have focused on verbal inter-action in mathematics lessons. Classroom data usually consist of either videotaped or audiotaped lessons that are transcribed to allow further analysis. The aim of this chapter is to discuss a methodology of transcript analysis influenced by Vygotskian ideas.

The problem of analyzing verbal interaction is assumed to be a major methodological problem for all research projects that focus on observing classroom processes. Even if it is not possible to expect general agreement on a methodology of analysis, a common attitude is assumed to be the rejection of standard schemes of discourse analysis. A critical review of the literature can be found in Edwards and Mercer (1987). These authors underline the fact that because standard schemes were devised to reveal linguistic structures, not educational or cognitive processes, they deal most explicitly with the form of what is said rather than with its content. In addition to the linguistic approaches, the same authors examine sociological and anthropological approaches that can give interesting insights into the social order of the classroom even though their scope does not focus on education.

The research project on Mathematical Discussion in Primary School was supported by the C.N.R. and the M.U.R.S.T. The project team consisted of the author and of the following teachers: M. Boni and F. Ferri (coordinators) and S. Anderlini, B. Betti, C. Costa, M. F. Ferraroni, C. Fortini, R. Garuti, M. L. Lapucci, S. Lombardini, F. Monari, A. Mucci, C. Ronchetti, S. Salvini, R. Tonolo, and M. P. Turchi.

A major problem of the sociological and anthropological approaches is presupposed to be the difficulty of taking into account the specificity of mathematics knowledge. Seeger (1991) criticized Voigt's (1985) interactional analysis because of its total shift of analytical attention from subject-matter structure to social-interactional structure and claimed that a theory of observation of classroom processes should include methods to analyze both the process of interaction and the object of teaching. Although the former may be borrowed, at least partially, from other research fields (such as social psychology and anthropology), the latter is assumed to be a genuine product of the research field of didactics of mathematics because of the specificity of mathematics knowledge.

However, a rigid distinction between the process of interaction and the object of teaching risks being misleading. Actually, what is the object of teaching? We cannot limit ourselves to content. We also have to include general attitudes toward mathematics that are evidenced, for instance, by participating in a mathematical discussion, asking mathematical questions, solving mathematical problems, proposing conjectures, listening to mathematical arguments, discussing mathematical models of reality, and so on. As attitudes change over time, they concern long-term goals. From a Vygotskian perspective, we could say that what has to be learned is determined by the *joint activity* between the teacher and pupils and that the joint activity comprises both the content features of the specific task and the quality of interaction.

The following discussion focuses on verbal interaction, a good detector of joint activity in the school setting because of the major importance of speech in formal instruction. This view follows the same trend represented by Kanes (chapter 6 in this volume), who claims that language constitutes instruction *and* the object of instruction. To allow further analysis, verbal interaction is referred to in the fixed form of annotated transcripts.

Is it possible to cope with the analysis of verbal interaction as a whole? The answer is negative if we allow only a single interpretation of transcripts. Even if every teaching episode is a unique event, it can be read in many different ways (Steinbring 1993), and every reading offers a different perspective.

The aim of this chapter is to present how interpretations of verbal interaction are given in a research project for didactical innovation on mathematical discussion (Bartolini Bussi 1991, 1992, 1996, in press a, in press b). Since every project for innovation aims at learning about classroom processes through designed transformations of classroom activity, the analysis mainly focuses on the coherence between the general motives (or long-term goals) of the teaching activity and the decisions of the teacher about the tasks as well as the communicative strategies adopted in classroom interaction. Because the analysis is highly dependent on basic assumptions, a brief description of the project is necessary to frame the presentation that follows.

THE MATHEMATICAL DISCUSSION PROJECT

The Mathematical Discussion Project has been in progress at Modena University since 1986. It is a project on didactical innovation in mathematics teaching developed by means of systematic cooperation between teachers and researchers. It focuses on long-term processes and general educational goals that concern advanced thinking and the construction of complex reasoning in compulsory education (grades 1–8).

Different models of research have been described elsewhere, and distinctions have been made among—

(a) analytical studies, in which a detached observer analyzes existing situations in standard classrooms to gain knowledge about classroom processes;

(b) action-research projects, in which innovation in school is produced without any consideration of (and sometimes even an ideological rejection of) the possibility of modeling the resulting processes;

(c) theoretical research, in which experiments in the classroom are introduced to build a coherent theory of the phenomena of mathematics teaching;

(d) research for innovation, in which innovation in school is both a means and a result of a progressive knowledge of classroom processes (for further details, see Bartolini Bussi [1994]).

The Mathematical Discussion Project is presently in the last category, so the poles of *action* in the classroom and of *knowledge* about classroom processes are developed in a dialectical relationship. The methodology of analyzing verbal interaction is developed inside the project in order to design (the pole of action) as well as to model classroom activity (the pole of knowledge) that is characterized by a systematic alternation of individual tasks and whole-class discussions. Teachers themselves are full members of the research group and are in charge of classroom observation as participant observers (Eisenhart 1988).

Mathematical discussion is a special kind of interaction that can take place in mathematics lessons. Mathematical discussion is conceived of as a polyphony of articulated voices on a mathematical object that is one of the motives for the teaching-learning activity (for further details, see Bartolini Bussi [1996, in press a]). The term *voice* is used in the sense of Bakhtin, after Wertsch (1991), to mean a form of speaking and thinking that represents the perspective of an individual (e.g., his or her conceptual horizon, his or her intention, and his or her view of the world) as a member of a particular social category. A form of mathematical discussion is the scientific debate that is introduced and orchestrated by the teacher on a common mathematical object to achieve a shared conclusion about the object of the debate. In this situation, the teacher utters a voice that represents the mathematical culture. The perspective on the object that is introduced by the teacher is usually different from the perspectives introduced by the pupils.

The emphasis on the different role of the teacher is in deep contrast to that of the models of peer interaction or of cooperative learning, which are preferred in most international research because of the presumed difficulty of reconciling the teacher's authority with the pupils' freedom to express their own ideas. The rationale for this choice lies in a conceptualization of the learning process that is deeply rooted in the European tradition and expressed by Vygotsky as follows: "Human learning presupposes a specific social nature and a process by which children grow into the intellectual life of those around them" (1978, p. 88). The famous genetic law of cultural development reads: "Any function in the child's cultural development appears twice, or on two planes. First it appears on the social plane, and then on the psychological plane. First it appears between people as an interpsychological category, and then within the child as an intrapsychological category" (Vygotsky 1981, p. 163). This is the almost universally quoted part that may be very easily related to the interpersonal activity that takes place in every discussion, from social chats to debates in scientific symposia. But the full quotation is longer and meaningful: "Social relations or relations among people genetically underlie all higher functions and their relationships. Hence one of the basic principles of volition is that of the division of functions among people, the new division into two parts of what is now combined into one. It is the development of a higher mental process in the drama that takes place among people" (Vygotsky ibid.). Hence, also in school settings, the division of functions is important. Even if teachers and students are engaged in the same indivisible activity (the teaching-learning activity in the school setting), their functions are not the same. The roles they play in the drama are different. The teacher is not one among peers, but rather is the guide in the metaphorical "zone of proximal development" (Vygotsky 1978). The process of learning is not so much determined by an individual relationship with knowledge but by the guidance provided by the adult members of society. The process of learning cannot be conceived of separately from the process of teaching, as is evidenced by the very word used throughout Vygotskian production, *Obuchenie*, which means "the process of transmission and appropriation of knowledge, capacities, abilities and methods of the human cognitive activity; it is a bilateral process, realized by teacher (prepodavanie, i.e., teaching) as well as learner (uchenie, i.e., learning)" (Menchiskaja [1966], as quoted by Mecacci in Vygotsky [1990]). To avoid identifying teaching with broadcasting ready-made knowledge, it is necessary to design (action) and study carefully (knowledge) exemplary processes in which pupils' appropriation of existing knowledge is not separate from their production of personal contributions. The Mathematical Discussion Project was expressly designed to meet this need.

To analyze long-term processes that cannot be reduced to the juxtaposition of short-term ones, a distinction is made, after Leont'ev (1978, 1981), among the levels of activity, actions, and operations. The first is the global level and corresponds to motives (the objects of teaching in the broad sense, including not only the mastery of mathematical concepts and procedures but also the attitudes toward mathematics). Activity is actualized by means of actions, defined by conscious goals, which include, for instance, the school tasks, solved individually or collectively; actions are realized by means of operations that directly depend on the conditions of attaining concrete goals.

Several kinds of analysis are made on the same teaching episode:

(*a*) Coarse-grained analysis, concerned with the levels of activity and actions and looking at the coherence between general motives and conscious goals of situations implemented in the classrooms

(*b*) Fine-grained analysis, concerned with all three levels—activity, actions, and operations—and looking at the coherence among general motives, conscious goals, and the teacher's operations to attain these goals

(c) Short-term analysis of effects, concerning the immediate effects of the teacher's operations on classroom processes

(d) Long-term analysis of effects, concerning the traces of motives in pupils' protocols

The following section contains an outline of the analysis of a teaching episode according to the foregoing scheme. The final remarks argue for the use of such an analysis to build new models of teaching activity.

EXAMPLE: THE ACTIVITY OF COORDINATING SPATIAL PERSPECTIVES AND THE CONCEPT OF POINT OF VIEW

Coarse-Grained Analysis: Location of the Action in the Global Activity

The example is taken from a teaching experiment on the coordination of spatial perspectives in first and second grades of primary school (Bartolini Bussi in press b). The general motive of the activity for pupils is to represent the global image of a part of the experiential world by organizing partial images (perspectives) obtained from different points of view. This motive has determined the features of a part of the classroom activity. The pupils have been confronted for some months with several tasks based on individual coding and decoding of some part of the visible world by means of different systems of signs (gesture, speech, drawing, written language), and they have participated in collective comparisons of their work.

At this point, the teacher proposes a discussion about the meaning of "point of view." What is at stake is the approach to the scientific concept of point of view. The focus on the term *point of view* is based on the Vygotskian hypothesis that in the early appropriation of a new word, the development of the corresponding concept has only just begun (Vygotsky 1990, ch. 6).

The aim of the discussion is to relate the meaning of the previous experiences, as it can be conceived of on the basis of an adult and decontextualized conception of point of view, to the personal senses given by the individual children to their specific drawings, utterances, and perceptions. The terms *meaning* and *personal sense* are used according to Leont'ev (1981). Meaning is crystallized and fixed in some medium (e.g., in a text such as "point of view is the fixed position from which somebody looks at some part of the visible world without moving the head") that describes the properties of the concept under scrutiny inside a system of theoretical knowledge (in this case, plane representation of three-dimensional space by means of linear perspective). Personal sense is the conscious sense created by each individual pupil to express the direction of his or her action. The explicit goal of the discussion for the teacher is precisely to encourage each pupil to express his or her personal sense through meaning.

Since meaning is to be expressed in a general form, the discussion has to be interpreted also as an introduction to the speech genre of formal instruction (Wertsch 1991, p. 112 ff.). The term *speech genre* is used in the sense of Bakhtin (as quoted in Wertsch 1991, p. 61): It is a typical form of utterance that corresponds to typical situations of speech communication, to typical themes, and, consequently, also to particular contact between the meaning of words and actual concrete reality under certain typical circumstances. This introduction is realized by the teacher by means of eliciting in the concrete and contextualized personal experiences the role and function of the general relation of things that could also apply to other cases.

The following excerpt examines a protocol to elicit the teacher's operations and to discuss their immediate effect and whether they are coherent with the motive and with the explicit goal of the discussion.

Fine-Grained Analysis of the Interaction

Contextual information

Seventeen first graders (ten girls and seven boys of very different learning levels) are sitting in a circle. The teacher is sitting among them or walking in the classroom. The situation is relaxed. The pupils are very young (6 to 7 years old), and, therefore, they sometimes need to be reassured by the teacher, not only with verbal utterances, but with gestures (a smile) or even physical contact (a hug). A tape recorder is in view, and everybody knows that activities with a tape recorder are important.

General features of the teacher's intervention

It is immediately evident that the verbal interaction in this episode is not regulated by the standard pattern of initiation-response-feedback

(Edwards and Mercer 1987). The teacher seldom intervenes and generally gives pupils the opportunity to express their ideas without immediate comments. She is also willing to accept, without anxiety, long pauses (up to a whole minute) to encourage pupils' assumption of responsibility and to underline the fact that reflection requires time.

If all the teacher's interventions are analyzed quantitatively, it can be said that their frequency is decreasing. Eleven interventions occurred in the first part (1–26). In the second, much longer and even more critical part (27–84), the teacher intervened only fifteen times, thereby allowing the discussion to flow by itself.

An epistemological analysis of the activity

Some main phases and subphases can be distinguished in the protocol. The phases are identified in relation to the epistemological analysis of the teaching activity.

Phase 1: From experiences to texts: the elicitation of personal senses

Subphase 1.1: (statements 1–4)

The question is, What does *point of view* mean? Pupils are asked a general question, which will be the leading theme of the whole discussion, but are allowed, in this phase, to construct texts that refer to their experience so they can link the personal sense of the previous problems to the term *point of view*. Whereas Ingrid tries immediately to give a general explanation, Elisa reconstructs a previous experience of a real-life drawing. Some days before, the pupils had drawn some objects (e.g., a bottle, a small box, a glass) that had been placed on the teacher's desk. Each pupil had drawn the object as it appeared from his or her own desk, so each pupil had a different perspective of the same objects. Claudio tries to reconstruct a personal experience, but the context of his story draws him elsewhere. He realizes it and is perplexed. This subphase is interrupted by a long pause (twenty seconds).

1. *Teacher:* Well, today I'd like to know something. It isn't something new but … I'd like to know if you have understood or understood better, if you know what does *point of view* mean. So many hands! Let's begin with Ingrid. Then make the rounds.

2. *Ingrid:* What to say, as if somebody who is on the right or on the left and looks at that thing, but the other one doesn't see it. The point of view is that one sees that thing and the other doesn't.

3. *Elisa:*   Mmm … one is on the right and he … one is on the right and the other is on the left. There is a thing between them: There is a bottle and there is a small box on the right and … there is a plastic glass on the left. Mmm … the one on the left cannot see the box on the right and the one on the right cannot see the glass if there is such a large thing in between.

4. *Claudio:* Like … like to have … like to have … somebody who has a small cat and he sees it on the left.… There is a child and the child wishes to go to the cat and the cat sees food and … (*perplexity*).

Silence (twenty seconds)

Subphase 1.2: (statements 5–10)

The question is repeated (5) to refocus attention on the task. Diego tries to reconstruct an out-of-school experience, where something is hidden by an interposed person, and Francesco refers to a previous school experience of a real-life drawing.

Up to Francesco's statement, nearly all the statements have a common opening: There's somebody on the right and somebody on the left. Because the teacher has implicitly accepted Ingrid's answer (2), the pupils presumably believe that this is the right format for an answer. Their attention is captured by the format, and what follows is not always pertinent or pertinently exposed, as if the contextual situation (two persons on different sides) has drawn stories into the classroom. The teacher (9) is struck by this formal regularity and tries to orient children elsewhere. However, she immediately realizes that her intervention could block Francesco's speech and corrects herself immediately.

5. *Teacher:*   I've asked what *point of view means*, do you remember the question: What does *point of view* mean?

6. *Diego:*   For instance … somebody ehm … ehm … ehm … ehm, somebody is on the right and somebody's on the left. They were walking and one saw something; the other went on and he didn't see well, he was old; the old man didn't see well and he crashed into it.

7. *Teacher:*   So, Diego, what does it mean, point of view?

8. *Francesco:* One is on the left and one is on the right.…

9. *Teacher:*   Again? OK, go on.

10. *Francesco:* The one on the left has a bottle, the one on the right has a glass. The one on the left, how can he see the glass if the bottle is taller?

Phase 2: From personal senses to meaning

Subphase 2.1: (statements 11–15)

The question is repeated (11) with an intentional shift toward general statements. Pupils are asked to express what *point of view* means without examples or maybe with other examples. The teacher is defining here the level of the expected answers, without inhibiting the production of other examples. Because the discussion has been carefully prepared, the teacher is aware that the goal of the discussion can be attained only by an explicit shift from the empirical level of everyday speech (where examples are offered, i.e., texts are built to describe one's own experiences) to the theoretical level of formal instruction (where texts are formulated in a decontextualized way, by means of generalizing the language so that it allows the text to be linked to other texts and to be applied in other situations). This shift creates a rupture with the previous flow of discussion, even if the teacher can hope to be helped by some pupils who have already generated some embryonic generalities in previous utterances. Gabriele (12) tries to answer by referring to a further example, but he does not succeed in completing the sentence, perhaps because he is aware that the answer is not of the expected kind. After a long pause (twenty seconds), the teacher asks again, "What did I ask?" This question (13), is different from questions (5) and (11). In the earlier instances, the teacher herself was controlling the pupils. In question (13), the teacher is aiming at fostering the pupil's own control on the pertinence of the answer. The intervention (15) is to correct an imprecise formulation.

11. *Teacher:* If the bottle is taller. Now, I'd like you to try and tell what does it mean, point of view, also without examples or maybe with other examples. As for you, what does it mean?

12. *Gabriele:* It means that you are somewhere and you look at a lamppost and then another and so...

Silence (twenty seconds)

13. *Teacher:* Yes. What did I ask, Gabriele?

14. *Gabriele:* What is point of view.

15. *Teacher:* What is, what does it mean?

Subphase 2.2: (statement 16)

Gabriele (a low-level pupil) constructs a statement that gives a general explanation. He introduces an important element, the need of a fixed eye. The teacher does not pick up on his suggestion.

16. *Gabriele:* It means that you are there, somewhere, you see a thing and then if you turn to the other side you see different things.

Subphase 2.3: (statements 17–22)

Francesca alludes to previous experiences of a real-life drawing. The teacher presses her to express herself better. The last statements (21–22) do not contain any reference to individual objects. Ingrid (22) explicates what Francesca (21) left implicit and introduces a new element, the observer's movement.

17. *Francesca:* The difference is that somebody sees a thing and the other sees a thing but a different one.

18. *Teacher:* I'm asking again....

19. *Francesca:* For instance, there is a table with a pen, a box, and a ball.... Somebody sees the pen and somebody sees the ball. We see different things.

20. *Teacher:* Different things.... I don't understand.

21. *Francesca:* From a point of view I don't see that which the other sees...

22. *Ingrid:* If the position changes.

Subphase 2.4: (statements 23–26)

The teacher poses an explicit question that induces the use of a more correct term and forces a verbal definition of *point of view*. Ingrid and Francesca together (24–25) construct a general statement that is repeated by the teacher (26) as a shared meaning for the classroom.

23. *Teacher:* How is the position named?

24. *Ingrid:* The point of view where we are...

25. *Francesca:* And we see.

26. *Teacher:* Pupils, pay attention. Ingrid and Francesca are saying that the point of view is the position, the point from where we observe.

Phase 3: The double movement

Subphase 3.1: (statements 27–35)

The teacher encourages the return to personal experiences by involving a low achiever in order to realize a double movement from experience to text and vice versa. New examples are introduced and accepted (35) by the teacher. The early format is overcome, and the discussion starts to flow without the teacher.

27. *Claudio (excited)*: I've understood!

28. *Teacher:*   I wonder if Claudio will come back to his cat story (*Claudio is perplexed*). OK, think it over, if you like.

29. *Stefano:*   [It means] that one cannot see the other thing.

Claudio and Gabriele are troubled.

30. *Teacher:*   No, boys, it is not OK. We cannot understand each other....

31. *Marco:*   One cannot see what he saw before.

32. *Elisa:*   I'll answer Marco. As bees see us enlarged, no, it does not work. Well, if I look at Ingrid from here, Ingrid, I can see your hair ...

33. *Francesco:*   But if I go there I see her face.

34. *Claudio:*   If I stay here I see the face, if I come there I see the hair.

35. *Teacher:*   Oh, from here I see the face and not the hair.

Subphase 3.2: (statements 36–38)

Elisa reconstructs, in a general way, the link between the position and the different views, and the teacher uses Elisa's statement (37) to recall the exact term. The insistence on the term *point of view* aims at relating it to pupils' personal senses.

36. *Elisa:*   The position is important.

37. *Teacher:*   This can be named....

38. *Chorus:*   Point of view.

Subphase 3.3: (statement 39)

The teacher recalls the pupils' statements with an explicit shift toward a general statement. The first part of the statement is nearly a literal coordination of pupils' statements, and the second part introduces the term *thing* in place of Ingrid and *parts* in place of either hair or face. Moreover, the teacher begins to introduce the pronoun *we* in the classroom discourse to underline the collective construction of this knowledge (pupils and teacher together).

39. *Teacher:* Before, you told me that if I change my point of view I see different things. Now we are saying that if I change the point of view of the "thing," somebody sees one part of the thing and somebody else sees another.

Silence (sixty seconds)

Subphase 3.4: (statements 40–43)

After a long pause (one minute), three pupils express, in their own words, the previous experience. Statement 43 is a repetition of statement 39, to accept previous statements.

40. *Gabriele:* It is possible to see another part of the object.
41. *Elena:* Like Elisa and Ingrid ... and Elisa sees only the back but if she goes to your place [to the teacher] she sees all the front.
42. *Greta:* It is actually true what we are telling now.
43. *Teacher:* Then, if we change the point of view, we see other things as well as other parts of the same object or person.

Subphase 3.5: (statements 44–46)

Stefano goes back to the personal experience of a real-life drawing. The teacher (45) forces him to use the right term, *point of view.*

44. *Stefano:* There is a glass, no, a pen-case too, wait a moment and ... on the other side there is Claudio and Claudio sees the glass and the pen-case behind and Diego has the pen-case in front, but if Diego goes where Claudio is, the position changes, Diego sees the position of Claudio.
45. *Teacher:* How did we say this exactly?
46. *Stefano:* Not the *position* but from his *point of view.*

Subphase 3.6: (statements 47–48)

Up to now, two different kinds of examples have been offered:

(*a*) Different points of view, with different observers in the same moment
(*b*) Different points of view, with the same observer in two different moments

Whereas the former refers to the static experience of viewing a real-life drawing from different positions, which pupils have done, the latter introduces a dynamic element, that is, the observer's movement. However, views depend on position only and not on the person. Elisa's statement is general; she repeats three times the term *everybody*. The teacher repeats her statement.

47. *Elisa:*   I wind up. If I go where you (the teacher) are I see the face but Elena sees the hair and if Elena comes with me, and Fabrizio comes and everybody comes with me where you are, everybody there.... Everybody sees Ingrid's face and Marco's face and nobody sees the hair.

48. *Teacher:*   Pay attention. She is saying that it is not because I am me, but everybody who comes here, where I am, sees her face.

Subphase 3.7: (statements 49–82)

An increasing number of pupils give their own statement about point of view. The teacher intervenes less and less. The levels of generality are different; some pupils speak about things and use impersonal pronouns, others speak about themselves and propose examples. Diego at last succeeds in giving an example.

49. *Francesco:* As one who sees the back of the schoolbag and one who sees the front. If the one who sees the back goes to the one who sees the front and the other goes in his place, the one who saw the back becomes the one who sees the front.

50. *Teacher:*   Diego?

51. *Diego:*   I've not raised my hand (*softly*).

52. *Teacher* (*with interest*): But you do have your point of view, don't you?

53. *Diego* (*aloud*): I do.

54. *Carmelina:* If you go there you don't see what you see now.

55. *Stefano:*   If you change your position you see other things or the same thing on the other side.

56. *Teacher:*   Does the object change if I change the point of view?

57. *Diego:*   For instance, if we.... Wait a moment till I think it over.

58. *Greta:*   If you change the point of view and there is the object, you see it from another point of view in another way.

59. *Elena:*   In front of the chalkboard there are four children who see the front of the four children who are close to the bookshelf; they don't see the back.

60. *Stefano:*   Sure, they turn their backs to the wall.

61. *Giusi:*   I agree with Elisa.

62. *Laura:*   Point of view means that one sees a thing, for instance a box, if he changes point of view and he saw the front side, later he sees the back side.

63. *Carmelina:* Like ... (*silence*)

64. *Greta:*   End up, please.

65. *Carmelina:* As if I were there and you there [two different places]. I change the point of view if I come there and I do not see the same thing any more and I see what you see.

66. *Diego:*   For instance ... no ... ouf.

67. *Greta:*   Changing the point of view, you do not see what you saw any more, you see in another way.

68. *Teacher:*   Good girl! Let's go on.

69. *Stefano:*   Changing the position, the point of view changes and the way of seeing things changes too.

70. *Laura:*   Point of ...

71. *Greta:*   ... view means ...

72. *Elisa:*   As in a drawing that is different for each of us, that is exactly what you see.

73. *Diego:*   Teacher, I am trying to speak. I can see from my point of view.

74. *Greta:*   Everybody sees from his point of view.

75. *Teacher:*   And things change according to the point of view.

76. *Diego*   (*with conviction*): I mean, I mean my brother and me, we have a desk, if I go left or right I see similar, but if I go in front of it, it's different. OK?

77. *Teacher:*   Good boy, you have succeeded in expressing your point of view.

78. *Fabrizio:*   Claudio can see my shoulder and if I come there I can see his shoulder.

79. *Diego:*   May I try again? I have a wardrobe ...

80. *Greta:*   End up please.

81. *Fabrizio:*   I'll end up myself. The point of view is important as we can see different things.

82. *Francesco:* It all depends. You can see also the same things.

Subphase 3.8: (statements 83–84)

Nobody in this discussion has yet recalled early shared experiences with landscapes, in which pupils had to describe verbally and draw a display (landscape) with a house, some puppets, and other objects (trees, cars, and so on). It had been the starting point of the whole activity; later real-life drawings of everyday objects had been substituted. The chorus states that the personal sense of that early experience can be expressed in the meaning of point of view.

83. *Teacher:* But did you like it?

84. *Chorus:* Yes, yes, the landscapes, the landscapes.

(A more detailed analysis can be found in Bartolini Bussi and Boni [1995]).

Short-Term Analysis of the Effects: Critical Remarks

Some effects of the teacher's intervention have already been described. To sum up, an observer might note a progressive shift from recalling particular experiences that focused on the different points of view assumed by two different observers, or by a moving observer, to stating the dependence of the point of view on the observer's position only. With only some exceptions, the teacher succeeds in orienting pupils toward general statements without inhibiting the expression of their thoughts. It is interesting to note that only two pupils, Katia and Luana, do not express themselves verbally, although their silent attention is monitored by the teacher.

Some problems of understanding seem to be solved in the course of the debate. Consider, for instance, the development of Diego, a low achiever, from (6) through (50–53), (57), (66), (73), to (76). After he has initial difficulty understanding the question, the teacher solicits his contribution; he tries unsuccessfully twice before finally giving an example. Something similar happens to Claudio, also a low achiever, even if we do not have enough utterances to monitor his development. Giusi, another low achiever, expresses only a generic agreement. They constitute, together with Fabrizio, the lowest achievement level in the classroom. The information collected in this discussion allowed the teacher to design some remedial activity for them to do in the following days.

A retrospective analysis of the discussion has made the teacher aware that in some cases she could have behaved differently. For instance, after (16) she could have picked up Gabriele's statements; in the second part of the discussion, she could have defended low achievers from Greta's exuberance. A further criticism concerned the lack of requiring each student to record an explicit, short, written statement in his or her notebook to store the result of the discussion for the future. This criticism was only methodological, since, in the situation under scrutiny, all the pupils remembered quite well the entire discussion (with a detailed reconstruction of the interventions) even four years later.

This teaching episode is a good example of the zone of proximal development (Vygotsky 1990) in a mathematical discussion. Some of the low-level pupils succeed in giving pertinent examples in the course of the discussion. Most of the pupils succeed in giving general statements—and not only a description of their own experience. So, joint activity with the teacher produces a performance that was not predictable for pupils alone; moreover, it could not even be solicited by pupils themselves, since general statements are typical of the speech genre of formal instruction in school setting.

On her part, the teacher succeeds in realizing the double movement from eliciting personal senses to meaning, from concrete to abstract, from contextualized to decontextualized (determined by adult knowledge of a theoretical meaning of *point of view*) and vice versa that is typical of theoretical thinking (Davydov 1990). Surely this attitude is not spontaneous in mathematics teachers, since they are inclined to move very quickly to, and to maintain, an abstract level even when some concrete experiences are proposed as starting points. Also, the difficulty this teacher had in alloting enough time to both aspects can be seen as she hurries to refocus on the general question in the first part of the discussion; later, the smooth flow of the discussion seems to reduce the teacher's anxiety.

Long-Term Analysis of the Effects

The discussion just described was carried out in June, at the end of the school year, so traces of long-term effects had to be looked for several months later.

In the second grade, no further clarification on the meaning of *point of view* was given. The term *point of view* was used again several months later when each pupil was given a cube, with an edge in the foreground, along with the following task: Draw the cube from your point of view exactly as you see it. The meaning of *point of view* proved to be among the control tools in the classroom, and the expression *point-of-view drawing* was repeatedly used by pupils in the individual protocols, and in the discussion, to mean drawing what you see and not what you know about the objects. Actually the development of a scientific concept of point of view continued to the fifth grade, through a teaching experiment on a plane representation of a three-dimensional world by means of perspective drawings (Bartolini Bussi 1996).

FINAL REMARKS

The last section presented the analysis of a typical kind of discussion that is systematically used in the Mathematical Discussion Project. It is a conceptualization discussion introduced by a question in a standard format: What does it mean ... ? What is ... ? (Bartolini Bussi 1991).

The foregoing discussions are related to the problem of definition in mathematics education and are assumed to be more and more important when an increasing number of mathematical concepts are introduced in the classroom as elements of a theoretical system. In our project, these discussions are usually located in teaching experiments after a series of school experiences (individual tasks, collective activities, or discussions) especially designed to offer pupils meaningful instances of the contextualized use of the concept under scrutiny. They are used prior to any formal definition that, in some instances, could be the outcome of the discussion itself. This has been our experience in many examples of teaching geometry in compulsory education (grades 1–8) (see, for instance, Mariotti [1994]).

A common feature of this kind of discussion, very evident from the epistemological analysis of the transcript, is the double movement from personal senses to meaning, and vice versa, that is encouraged by the teacher. It is a kind of guiding plan that has been assumed by the teacher in the designing phase to allot time for eliciting personal senses and the shared meaning.

The construction of meaning is the crucial problem of mathematics education. On the one hand, no meaning can be taught directly, since every explicit statement of a definition could be memorized by pupils and repeated to meet the teacher's expectations, without being linked to any previous experience and without being applicable to any further problem. On the other hand, no meaning can be the matter of negotiation, since scientific concepts are not to be created anew in school (Vygotsky 1990) but are to be assimilated as the products of centuries of development by humankind. This situation generates a well-known paradox. We believe that a part of the solution of this paradox is in creating and supporting the double movement between personal senses and meanings. Developing a better and better definition of the teacher's role in specific situations seems to be necessary for creating new models of teaching, which is actually the aim of the analysis developed in the Mathematical Discussion Project.

REFERENCES

Bartolini Bussi, M. G. "Social Interaction and Mathematical Knowledge." In *Proceedings of the Fifteenth Conference of the International Group for the Psychology of Mathematics Education*, edited by F. Furlinghetti, pp. 1–16. Assisi, Italy: International Group for the Psychology of Mathematics Education, 1991.

———. "Mathematics Knowledge as a Collective Enterprise." In *The Dialogue between Theory and Practice in Mathematics Education: Overcoming the Broadcast Metaphor: Proceedings of the Fourth Conference on Systematic Cooperation between Theory and Practice in Mathematics Education*, 1990, edited by F. Seeger and H. Steinbring, pp. 121–51. Materialen und Studien des IDM, Heft 38. Bielefeld: Institut für Didaktik der Mathematik, 1992.

———. "Theoretical and Empirical Approaches to Classroom Interaction." In *Didactics of Mathematics as a Scientific Discipline*, edited by R. Biehler, R. W. Scholz, R. Strässer, and B. Winkelmann, pp. 121–32. Dordrecht, Netherlands: Kluwer Academic Publishers, 1994.

———. "Mathematical Discussion and Perspective Drawing in Primary School." *Educational Studies in Mathematics* 31 (1996): 11–41.

———. "Joint Activity in Mathematics Classrooms: A Vygotskian Perspective." In *The Culture of the Mathematics Classroom: Analyses and Changes*, edited by F. Seeger, J. Voigt, and U. Waschescio, pp. 13–49. Cambridge: Cambridge University Press, in press a.

———. "Coordination of Spatial Perspectives: An Illustrative Example of Internalization of Strategies in Real-Life Drawing." *Journal of Mathematical Behavior*, in press b.

Bartolini Bussi, M. G., and M. Boni. "Analisi dell'interazione verbale nella discussione matematica: un approccio vygotskiano." *L'insegnamento della matematica e delle scienze integrate* 18 (1995): 221–56.

Davydov, V. V. *Types of Generalization in Instruction.* Soviet Studies in Mathematics Education, vol. 2. Reston, Va.: National Council of Teachers of Mathematics, 1990.

Edwards, D., and N. Mercer. *Common Knowledge: The Development of Understanding in the Classroom.* London: Routledge, 1987.

Eisenhart, M. A. "The Ethnographic Research Tradition and Mathematics Education Research." *Journal for Research in Mathematics Education* 19 (1988): 99–114.

Leont'ev, A. N. *Activity, Consciousness and Personality.* Englewood Cliffs, N.J.: Prentice Hall, 1978.

————. *Problems in the Development of Mind.* Moscow: Progress, 1981.

Mariotti, M. A. "Figural and Conceptual Aspects in a Defining Process." In *Proceedings of the Eighteenth Conference of the International Group for the Psychology of Mathematics Education,* edited by J. P. da Ponte and J. F. Matos, pp. 232–38. Lisbon: Program Committee of the Eighteenth Conference of the International Group for the Psychology of Mathematics Education, 1994.

Seeger, F. "Interaction and Knowledge in Mathematics Teaching." *Recherches en didactique des mathématiques* 11 (1991): 125–66.

Steinbring, H. "Problems in the Development of Mathematical Knowledge in the Classroom: The Case of a Calculus Lesson." *For the Learning of Mathematics* 13 (1993): 37–50.

Voigt, J. "Patterns and Routines in Classroom Interaction." *Recherches en didactique des mathématiques* 6 (1985): 69–118.

Vygotsky, L. S. *Mind in Society: The Development of Higher Psychological Processes.* Cambridge: Harvard University Press, 1978.

————. *Pensiero e linguaggio.* Critical edition by L. Mecacci. Bari: Laterza, 1990.

Wertsch, J. *Voices of the Mind: A Sociocultural Approach to Mediated Action.* Cambridge: Harvard University Press, 1991.

4

Discourse and Beyond: On the Ethnography of Classroom Discourse

Falk Seeger
University of Bielefeld

> I distrust summaries, any kind of gliding through time, any too great a claim that one is in control of what one recounts; I think someone who claims to understand but who is obviously calm, someone who claims to write with emotion recollected in tranquillity, is a fool and a liar. To understand is to tremble. To recollect is to re-enter and be river. An acrobat after spinning through the air in a mockery of flight stands erect on his perch and mockingly takes his bow as if what he is being applauded for was easy for him and cost him nothing, although meanwhile he is covered with sweat and his smile is edged with a relief chilling to think about; he is indulging in a show-business style; he is pretending to be superhuman. I am bored with that and with where it has brought us. I admire the authority of being on one's knees in front of the event.
>
> —Harold Brodkey, *Stories in an Almost Classical Mode*

THE present chapter attempts to discuss two basically conflicting ideas: the idea that classroom discourse is a principal target area for research in teaching and learning mathematics and the idea that giving a complete picture of teaching and learning mathematics may require more than reconstructing it as discourse. The perspective that in the end will serve as a basis to reconcile these conflicting ideas is that teaching and learning are understood as the introduction into, and the sharing of, a culture. Here, the relation between the content and the forms of a culture will play an important role, as well as the notion that culture appears as a multiple, hybrid concept.

85

When I was reflecting on the problem of analyzing discourse in mathematics classrooms, an interesting analogy with a certain cookbook came to mind. In the cookbook's introduction, the author tried to lay some educational groundwork by giving the reader reasons for not being ashamed to enjoy the sensuous delights of preparing and eating food. Quoting from a famous journal on the art of cooking, he tried to show that in the past, people knew how to enjoy a dinner. He reprinted the menu of a dinner that had actually been served, comprising twelve courses for a total price of 100 marks (without the wine).

The menu was impressive in terms of the sheer quantity of food, its diversity, and the choice of wines, but the impression remained rather opaque for the reader trying to imagine how the dinner was organized and what kind of behavior people showed or had to show. The whole meaning of a dinner, as part of the culture, could not be read from the menu.

It seems that it is equally hard, or even impossible, to grasp the meaning of being introduced into the culture of mathematics by reading the "content" of classroom discourse (let alone the curricula) as it is to get an idea of a dinner as a cultural phenomenon by reading its menu. It is a truism "that what counts most in a culture is what people do" (Goodnow 1993, pp. 372–73). But this may not be the whole truth; it also counts what people *are*.

The first section of this chapter briefly describes a framework of a theory of learning as introduction into a culture. The second section presents an analysis of form versus content in classroom discourse, and the third, concluding section reflects on the consequences for classroom research in mathematics education.

LEARNING THE CULTURE AND THE CULTURE OF LEARNING

The relation between learning and culture has attracted considerable attention in the recent past, two plausible reasons being the deeply felt need to change classroom culture and the pressure from changes in the culture at large. In mathematics education, constructivist and Vygotskian approaches are seeking alternatives to traditional approaches (see, e.g., Cestari92
and Bartolini Bussi, chapters 8 and 3, respectively, in this volume).

Both approaches can be characterized as "transactional." Different emphasis is laid, however, on the relation between the individual and the interindividual. Vygotskian approaches usually assume that the interindividual plays an outstanding role in the development of the individual where "culture" is seen as the dominant interindividual force (see, e.g., Bruner [1985]; Raeithel [1990]; Vygotsky [1981, 1987]; Wertsch [1985, 1991]).

Bruner describes what part culture is to play in a theory of education. Here, Bruner characterizes how his own view on learning changed from his first publications on learning and education in the 1960s:

> My model of the child in those days was very much in the tradition of the solo child mastering the world by representing it to himself in his own terms. In the intervening years I have come increasingly to recognize that most learning in most settings is a communal activity, a sharing of the culture. It is not just that the child must make his knowledge his own, but that he must make it his own in community of those who share his sense of belonging to a culture. It is this that leads me to emphasize not only discovery and invention but the importance of negotiating and sharing— in a word, of joint culture creating as an object of schooling and as an appropriate step en route to becoming a member of the adult society in which one lives out one's life. (Bruner 1986, p. 127)

We find here a fundamental distinction between two modes of learning:

- Learning as interaction between subject and object
- Learning as interaction between subjects

Bruner (1986) also distinguishes between two major modes of thought—the narrative and the paradigmatic—which roughly correspond to the difference between oral skills and logical-scientific ones. How are these two modes of learning related? Since the time of Piaget, the first mode has been seen as basic, whereas the second has been viewed as providing only an additional dimension to the subject-object interaction—the learning process is only made more complex if the social dimension is considered. Basically, knowledge arises at the interface between subject and object, as has been demonstrated by Piaget's numerous studies (see, e.g., Piaget [1976]). The introduction of the notion of learning as the sharing of a culture, however, does not simply add another element to the interaction of subject and object—it changes the whole scene. It is implicit in this notion that our encounters with the real world of objects are mostly not direct but mediated by tools or signs, as Vygotsky (1978) had stressed.

But tools and signs do not mediate only between subject and object, they also mediate the cultural forms of the practical life of humans. By acquiring the forms of how to use these "psychological" means, a child gets introduced into a culture. It has been advocated that the concept of the schema, or schematic anticipation, has considerable explanatory power (Selz 1924; see also Piaget's discussion of Selz [1961]).

Thus, sharing a culture and getting introduced into a culture happen as children make the schemata or forms of that culture a content of their activity. Davydov and Zinchenko have described the process in the following way:

> Forms of culture are steps in an already ordained development of consciousness of the individual. One can say that this raises a fundamental problem for science, namely, to determine how the content of the spiritual development of mankind is converted into its forms and the assimilation of these forms by the individual becomes the content of the development of his consciousness. (Davydov and Zinchenko 1981, p. 41)

This excerpt aptly expresses the idea that being introduced into and sharing a culture cannot be based on a simple transmission of knowledge from the adult to the child, the expert to the novice, the teacher to the pupil. The process is much more complicated, and its understanding makes it necessary to scrutinize the interplay between cultural form and content.

Up to this point, when "culture" was talked about, what the concept should signify was left unspecified. Elias (1978) has stressed that it is a concept that expresses the self-consciousness of the West and its alleged superiority over less developed countries in terms of the level of its technology, the nature of its manners, its scientific knowledge, its world view, and much more. Even though it would be interesting to follow this line of thought, for example, under the perspective of the recent interest in ethnomathematics, it will not be discussed here. Instead, the discussion will focus on the necessity to take apart that large concept. Culture, obviously, is not a single-layer phenomenon, a unitary concept. It has many layers and many levels in which processes can be found that work according to different principles.

Getting introduced into a culture, thus, cannot be fully understood by looking only at the "content" of that culture that is distributed and acquired. Getting introduced into a culture means to make the forms, the manners, one's own; it affords certain cognitive structures in which those structures in turn produce culture. In a fascinating book on the evolution of culture and cognition, Donald (1991) has presented a conceptual framework that illuminates culture and the modern mind as a hybrid phenomenon. Domain-specificity has become here a highly general principle of development, since different domains of culture evolve at different periods in human evolution. The story of the evolution of human culture and cognition, basically, comprises three major transitions: the transition from episodic to mimetic, from mimetic to mythic, and from mythic to theoretic culture. It begins with episodic culture and cognition, which humans still share with their primate ancestors.

> From a human viewpoint, the limitations of episodic culture are in the realm of representation. Animals excel at situational analysis and recall but cannot represent a situation to reflect on it, either individually or collectively. This is a serious memory limitation; there is no equivalent of semantic structure in an animal memory, despite the presence of a great deal of situational knowledge. Semantic memory depends on the existence of abstract, distinctively human representational systems. The cognitive evolution of human culture is, on one level, largely the story of the development of various semantic representational systems. (Donald 1991, p. 160)

In the transition from episodic to mimetic culture, a new level of cultural cognition and representation evolved. "Mimetic skill or mimesis rests on the ability to produce conscious, self-initiated, representational acts that are intentional but not linguistic" (Donald 1991, p. 168).

If we take children's play as an example of the rich occurrence of mimetic skill, it becomes evident that the mimetic is still part and parcel of today's culture, as well as an outstanding route to acquire its basics. In the evolution of culture, mimetic skill sets the stage for the evolution of language, since it presupposed and developed a number of representational features that are typical for language, such as generativity and the separation of representation and referent. Social consequences of the mimetic skill were important, too, because it provided a basis for sharing a collective conceptual model of social structure, for reciprocal social relationships, and for the coordination of group activities. The child's acculturation into a mimetic society required some form of pedagogy—and with that, also some basic insight into the limits of the tutee, some basic notion of a "zone of proximal development," and some ability to sense on the side of the adult what the young can, and cannot yet, learn.

A second transition is from mimetic to oral, narrative culture, called mythic by Donald because myth appears as the dominant feature of the cognitive organization at that cultural level. Mythic culture created a huge leap forward toward integrating and making coherent the scattered knowledge of the world. However, the cognitive organization it afforded remained largely one that put a heavy load on internal memory because the narratives had to be remembered verbatim (cf. Havelock 1986).

In the final transition from mythic to theoretic culture, the dominance of spoken language and narrative style of thought was broken on the basis of three crucial cognitive phenomena that became the hallmark of theoretic culture: graphic invention, external memory, and theory construction. Alphabetic writing, print, and computers are all examples of theoretic culture that allow the external storage and retrieval of accumulated knowledge and experience. From a psychological perspective, the analysis of external memory proves to be difficult, since there are no ready theoretical concepts to grasp the specificity of external memory. Only an analogy to the concept of network might be helpful. Thus, a number of current approaches to distributed cognition (cf., e.g., Hutchins [1991]) use the heuristic power of the network concept.

Even though the feasibility of condensing the treatment of an embracing approach in only a few lines is uncertain, the foregoing presentation of Donald's framework should have illustrated that culture and cognition are hybrid phenomena and that language and speech are but one, albeit important, aspect of culture. One major conclusion is that learning has to be seen as being different in different levels of culture. The cognitive consequences and results in mimetic culture are different from those of oral, narrative culture, and both differ from those of theoretic culture.

It can be concluded, therefore, that there is some heuristic basis to the assumption that classroom learning is best understood as a hybridization of three types of learning:

- Mimetic learning, based on observing and performing gestures and the like; basically a type of learning that Lave and Wenger (1991) have termed *legitimate peripheral participation*
- Discursive learning, based on speech and narrative, including such phenomena as metaphors, speech genres (Halliday 1978; Lemke 1988), and myths (Seeger and Steinbring 1994)
- Theoretic learning, based on the externalization of knowledge, including literacy, print, and computers. Theoretic learning creates and uses relational thinking as opposed to substantial thinking (cf. Cassirer 1955). Its basic orientation is a fundamental demythologization.

If it is correct to say that learning the culture is in fact a problem of learning different types of culture in school, then it is also correct to say that it can start from different entry levels, which is only another expression for the idea that the relation between form and content, obviously, is different in learning the different cultures.

FORM AND CONTENT
IN CLASSROOM DISCOURSE

Although the previous section underlined the multilevel nature of learning, this section focuses exclusively on classroom discourse, partly because the foregoing theoretical part came only after the empirical study was completed. Only after the transcripts were analyzed did it become increasingly necessary to formulate a framework that could take into account an apparent contradiction between the impression of the respective discourses on the spot and the results of the analyses of the transcripts of that same lesson.

Investigating the forms of verbal interaction and classroom discourse requires dealing with a certain pattern or order of discursive processes. Mostly, of course, these patterns are recurrent. Recurrence of patterns expresses some underlying functional necessity. At the surface level, however, is the impression of a certain order of classroom discourse, or a certain disorder.

The following section contains a comparison between examples of teaching episodes that were taken from two teachers in a research project conducted with Ute Waschescio and Heinz Steinbring at the Institut für Didaktik der Mathematik in Bielefeld. For reasons of anonymity and convenience, the teachers are called Mrs. L. and Mrs. T. Both teachers agreed to take part in testing "open material" for students with learning disabilities, which was an objective of the project. They were offered a series of units from the open material, from which they picked what they liked. In addition, the researchers' intentions were discussed with them. They were also given a short introductory text containing the overall philosophy of the material, which emphasized the importance of relational knowledge in mathematics and the problems of relating mathematical knowledge and procedures to concrete experience. The introductory text also emphasized that it was not enough to center mathematical activity on manipulatives, routines, and actions. Reflecting on what one was doing was put forward as an important aspect of mathematical activity, which underlined the necessity of discussing and exchanging mathematical ideas in class.

Thus, in these situations, the teachers had chosen—for whatever reasons—some units they wanted to test in their classes.

At first sight, the teaching of Mrs. L. simply appeared better in comparison with Mrs. T.'s teaching. Mrs. L.'s teaching appeared well structured, very systematic, and oriented toward definite results. There were no "loose ends"; in each lesson the goal that she set up for that lesson was attained. In addition, her behavior was very genuine. She always showed the right amount of affect when she praised or criticized a student.

The quality of Mrs. L.'s teaching, at a closer look, can be characterized by a certain kind of teacher-student dialogue. This kind of dialogue pressed for a certain result by using questions and answers, alternating teacher question, student answer, teacher question, and so on.

The following sequence from a lesson transcript illustrates this well-known kind of dialogue. This sequence is about calculating the average from a series of five values that had been collected in an experiment on personal reaction time. Five values are written on the chalkboard and their sum is written below.

619. *Mrs. L.:* Now, the question arises, which is written here below, how do I calculate the average? ... We have once already ...

620. *Suria:* Divided.

621. *Mrs. L.:* Calculated the average. There are now five attempts. 1, 2, 3, 4, 5. And this is the result of all five attempts added together.

622. *Suria:* Dividing.

623. *Mrs. L.:* Milan, are you paying attention?

624. *Özlem:* Here!

625. *Mrs. L.:* Now, how do we find that out? And wha ...

626. *Özlem:* Here!

627. *Mrs. L.:* Yes, Özlem?

628. *Özlem:* Divided by two.

629. *Mrs. L.:* Divided by two, you mean, is the average? Why by two?

630. *Özlem:* Ahem, don't know myself.

631. *Mrs. L.:* Another suggestion?

632. *Suria:* The half.

633. *Mrs. L.:* Divided by two would be half. Do I want to know what is half?

634. *Suria:* No, the average.

635. *Several students simultaneously:* No.

636. *Mrs. L.:* Look again. This is a particularly good result.

(Mrs L. points to the first value, 2, at the chalkboard.)

637. *Mrs. L.:* And which number is the particularly bad result?

638. *Suria:* 30.

639. *Mrs. L.:* That is number 5. So, I want to know the average. What now is the average?

In this episode the teacher tries without success to call back to students' memory the procedure of how to calculate the average. At the end of the episode, the same question is put as at its beginning; it has become evident only that average cannot be the half. This episode illustrates that strictly following the pattern of taking turns—by the teacher (question) and the student (answer)—was not a successful repair strategy for the situation at hand. The students should remember how they once did calculate the average. But the attempt to activate the collective social memory of the class failed. One reason was that this type of educational discourse does not give students room to articulate their ideas and concepts. Students are granted the possibility to fill in only the slots that are left by the teacher's questions—with the right answer. This episode is only a short example of the style of communication that this teacher used throughout the lessons, especially in the phases where a result should be reached. This kind of educational discourse has other characteristics: all student utterances are teacher initiated; virtually no verbal interaction takes place among students; there are virtually no student questions; students' utterances consist mostly of short sentences or just one word; no discussion occurs among students or between students and teacher.

Because the only thing that counts in this kind of discourse is the right answer, the searching movements of the students have much in common with the random movement of water molecules in a *laminar* flow. But even though the movements of the elements are ruled by chance, at the surface the picture of a perfectly ordered flow emerges.

In contrast to the impression of order, Mrs. T.'s teaching and class at first sight appeared chaotic. A high degree of unrest among students was common and was produced by their many verbal interactions, only some of which belonged to the subject of teaching. The following episode, which illustrates the *turbulent flow* of events, is about a game called "take 33."

"Take 33" is a game for two persons. From 33 they subtract alternately 3 or 5 until 2, 1, or 0 is reached. A 2 means that the student who played second has won; a 1, that the student who played first has won; and a 0, that a draw occurs. The diagram in figure 4.1 is used in playing the game.

The episode in question is about noting on the chalkboard the results of the games the students had been playing. The teacher's intention is

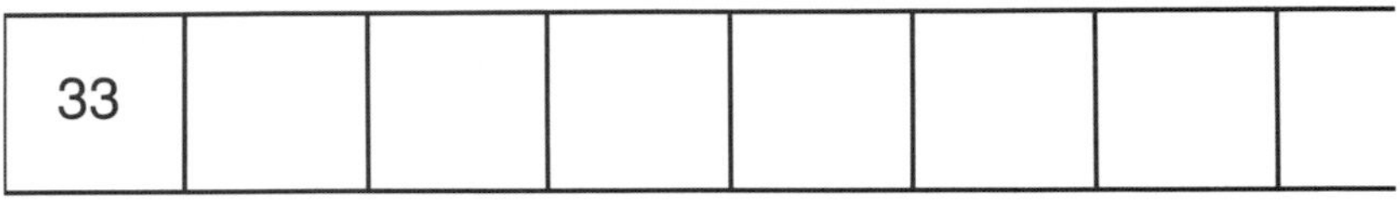

33							

Fig. 4.1

to arrive at some generalization about the best strategy to win the game. The result of a particular group, Gerd and Andy, is reported by Gerd and then written on the chalkboard.

907. *Mrs. T.:* So, we're just waiting for Frank.... I should also like to go on, it was a bit loud just now.

908. *Frank:* Yes, do go on.

909. *Mrs. T.:* Gerd, tell me the, uh, I've heard you've obtained one time one.

910. *Gerd:* Yes.

911. *Mrs. T.:* Who has begun here?

912. *Gerd:* Me.

913. *Mrs. T.:* So, then you had also won. Now tell me which numbers you have noted.

914. *Gerd:* Minus 3.

(Teacher writes "30" inside and "–3" on top of the first slot.)

915. *Student:* That is won.

916. *Student:* ... has won again.

917. *Mrs. T.:* And then? Go on, Gerd!

918. *Gerd:* Minus 5, always alternately.

(Teacher writes "25" inside and "–5" on top of the second slot.)

919. *Student:* I am sorry, I've also minus 1.

920. *Mrs. T.:* Ah yes!

921. *Student:* Well, no wonder, if they always do it alternately. Just 3, 5

922. *Mrs. T.:* Always beginning ...

923. *Gerd:* Always alternately. We tried it out. That [unintelligible]

(Mrs. T. finishes the series, with "25," "22," "17," "14," "9," "6," and "1.")

(Disorder.)

924. *Student:* Three, five, three, five, three, five, pah, eh!

925. *Ina:* Three times draw and once won.

926. *Frank:* Who has three draws?

927. *Ina:* It's her.

928. *Student:* We've got ...

929. *Mrs. T.:* Yes, how?

930. *Michael:* [unintelligible]

931. *Gerd:* Him? 17?

932. *Student:* Yes, 14.

933. *Student:* Got no chance.

934. *Gerd:* 9, 6, 1.

(Pause)

935. *Mrs. T.:* So, now these are.... How many times did you subtract 5?

936. *Student:* Swell!

937. *Gerd:* Not me, Andy.

938. *Student:* Me.

939. *Mrs. T.:* Andy.

940. *Andy:* 4 times. Don't know ... 4 times.... Show me.

941. *Student:* Yes, 4.

942. *Andy:* 1, 2, 3, 4 times.

(Teacher writes "4 times 5" on top of the chalkboard.)

943. *Mrs. T.:* And how many times did you subtract 3?

944. *Student:* 4.

(Teacher writes "4 times 3.")

945. *Student:* 12.

This sequence nicely illustrates the different streams of communication that ran parallel in that lesson. Aside from the official theme, writing the results on the chalkboard, the students exchanged their results. While the teacher was working with one student on the official task, different groups of students were working on unofficial tasks and exchanging their results. Attention was continuously fluctuating between the chalkboard and the peer students, which gave the impression of chaos and confusion. In addition, the teacher at some points was not sure whether the calculations were correct. Generally, the activities of the students were not directed toward finding strategies to win; they first of all wanted to know who had won or how many draws had occurred. But in contrast to the class of Mrs. L., the students here articulated their own point of view, albeit in the form of the unofficial discourse and with a taste for chaos and confusion.

These two examples underline the importance of the *forms* of classroom discourse. Content seems to be, by and large, rather irrelevant. It seems only to serve the point of view of the observer qualifying or disqualifying the attempts of teachers and students as successful or not, as mathematically or didactically sound or not, and so on.

It can be seen that the traditional order of classroom discourse produces patterns that seem to follow random distribution, whereas the turbulent patterns exhibit more order at least as far as the point of view of the students is concerned. This observation underlines the assumptions that confusion is not necessarily detrimental to learning and that order does not necessarily enhance learning (cf. Brown 1992).

It can also be seen that what can be viewed as "taken as shared" requires some knowledge about how students articulate their perspectives in classroom discourse.

Research in classrooms has documented again and again that classroom discourse lacks one of the outstanding features of a normal conversation, that is, the reciprocity of perspectives as it appears in taking turns. The situation is well described by Edwards and Furlong:

> In any piece of social interaction, it may become apparent that the participants do not have a reciprocity of perspectives. In the course of conversation, they may become aware that each means something different by a key term. If the difference is too obtrusive, it becomes a stumbling-block to the conversation.... In talk between equals, neither has the right to insist on *his* definition or the obligation to wait for a ruling. But in most classrooms, academic meanings are the province of the teacher. The pupil will normally suspend any knowledge he has about the subject until he has found out the teacher's frame of reference, and moved (or appeared to move) into it. For the academic curriculum to proceed, a reciprocity of meanings has to be established. But in the unequal relationship between most teachers and pupils, the movement is nearly always in one direction; the pupil has to step into the teacher's system of meanings and leave them relatively undisturbed. Being taught usually means suspending your own interpretations of the subject matter and searching out what the teacher means.... The pupil's suspension of his own interpretation may be so complete that if he cannot understand what the material means to the teacher, then it becomes literally meaningless for himself. (Edwards and Furlong 1978, p. 104)

Thus, classroom discourse is far from negotiating meaning. "Negotiating" suggests an interaction between equals. Negotiating meaning seems more like a program for the future classroom than an empirical finding.

The question of whether students in the classroom play the "language game" of mathematics (cf. Kanes, chapter 6 in this volume) may be a question not only of power or of classroom norms. It is also a question of skill. As Lemke (1988) has pointed out, a specific speech genre must be learned for each subject in school. Speech genres in classroom education uniquely blend certain forms of discourse with the specific content of the subject.

> The mastery of every subject is a mastery of the semantic patterns of some particular thematic formations and of the genre structures of the subject and their uses. (Lemke 1988, p. 98)

What creates enormous problems for students is that the speech genre specific to a certain subject is not explicitly taught or even talked about. One consequence is that the different aspects of the speech genre of mathematics in schools—discussion, argumentation, debate, monologue, dialogue, and so on—have to be explicitly treated in the classroom.

THEORY AND PRACTICE:
THE ALIEN AND THE FAMILIAR

I assume that the situation of a researcher can best be described as an ethnographic perspective. The ethnographic perspective is typically taken by an observer who tries to understand cultural phenomena that appear alien to her or him. A fundamental problem can be seen in the tendency to impose the world view of the observer on the alien world view. In ethnography, many reasons can be found for this tendency: alien cultures are principally understood as inferior to the Western world view and as less rational, and their practices are understood as unnecessary and complicated by strange magical rules, prohibitions, and prescriptions. An extensive discussion is necessary to make clear that it is an extraordinary task to describe and understand a culture in its own terms (see, e.g., Sahlins [1976]; Habermas [1984]).

Noteworthy parallels can be found between the situation of a cultural anthropologist or ethnographer who tries to understand an alien culture and that of the researcher in mathematics education who tries to understand communicative processes in the classroom. One of the major impediments to understanding seems to be that the researcher usually knows a lot of mathematics that can prevent her or him from fully understanding what goes on in its own terms. "Making the familiar strange," a slogan of ethnographic research in a familiar culture, is hard to accomplish if the main attempt is directed toward making the strange and complex web of processes familiar by projecting it onto the frame of reference of mathematics.

The impression one sometimes gets while reading interpretations of observations from mathematics classrooms is that the researchers often find it hard to make this whole chaos and confusion compatible with the transparent mathematical content. It seems that they often deplore that mathematical content is exposed to the jungle of the classroom where—as the results indicate again and again—it has no chance to survive; the dynamics of social interaction dominate and stifle mathematical content. This notion—admittedly somewhat exaggerated—seems to be the consequence of an emphasis laid on mathematical content that makes it hard in the end to understand the diversity of perspectives held by teacher and students. "The ethnography of thinking … is an attempt not to exalt diversity but to take it seriously as itself an object of analytic description and interpretive reflection" (Geertz 1983, p. 154).

FINAL REMARKS

Doing research on learning in the classroom, from the ethnographic perspective of sharing a culture, is primarily focused on discourse. Language and speech form the exterior plane from which the internalized forms of thinking develop. If the didactic voice (Edwards and Furlong 1978) today primarily articulates the perspective of the teacher and leaves not much room for the perspective of the student, then it is small wonder that introducing children to the culture of mathematics is not as successful as it should be. Taking stance and counterstance does not belong to the usual form of classroom discourse that one meets in an ordinary mathematics classroom. What seems necessary in this situation is an attempt to better understand the rationality of the prevailing classroom discourse that underlies the recurrent patterns. To reflect on that, and on the possibility of change, is not a task that has to be approached separately by teachers and researchers (cf. Bartolini Bussi [1994], and chapter 3 in this volume). Developing the dialogue is simply a consequence of applying what has been said.

We are faced with the difficult situation that the processes of collective argumentation, of taking stance and counterstance, of discussion are hardly to be found in ordinary mathematics classrooms (for counterexamples see Bartolini Bussi [1992, 1994] and chapter 3 in this volume). We seldom find an open context where the perspectivity of mathematical activity—the fact that something is said from a certain point of view—has become a content. In a certain sense we find that realizing the circumstances of the classroom and changing them coincide: for research to be done in the nature of the phenomena described, they have to become real in the actions of teachers and students in the first place. Since the time of Karl Marx, we have known that talk about the coincidence of realizing and changing circumstances is talk about a revolutionary situation.

The seeds for the current revolutionary situation in the culture of schools and learning were planted long ago with the beginning externalization of knowledge and the dawn of theoretic culture. Theoretic culture also changes the meaning of mimetic and discursive culture. It is a challenge for the future culture of learning that mimetic learning (e.g., how to use a mathematical tool) and discursive learning (e.g., how to argue collectively about mathematics) have to gain a changed meaning under the influence of theoretic culture that has changed everything.

REFERENCES

Bartolini Bussi, M. G. "Mathematics Knowledge as a Collective Enterprise." In *The Dialogue between Theory and Practice in Mathematics Education: Overcoming the Broadcast Metaphor: Proceedings of the Fourth Conference on Systematic Cooperation between Theory and Practice in Mathematics Education, Brakel, Germany, 1990*, edited by F. Seeger and H. Steinbring, pp. 121 51. Materialen und Studien des IDM, Heft 38. Bielefeld: Institut für Didaktik der Mathematik, 1992.

———. "Theoretical and Empirical Approaches to Classroom Interaction." In *Didactics of Mathematics as a Scientific Discipline*, edited by R. Biehler, R. W. Scholz, R. Strässer, and B. Winkelmann, pp. 121–32. Dordrecht, Netherlands: Kluwer Academic Publishers, 1994.

Brown, S. I. "How Much Static Can the Broadcast Metaphor Stand? Towards a Pedagogy of Confusion." In *The Dialogue between Theory and Practice in Mathematics Education: Overcoming the Broadcast Metaphor: Proceedings of the Fourth Conference on Systematic Cooperation between Theory and Practice in Mathematics Education, Brakel. Germany, 1990*, edited by F. Seeger and H. Steinbring, pp. 7–32. Materialen und Studien des IDM, Heft 38. Bielefeld: Institut für Didaktik der Mathematik, 1992.

Bruner, J. S. *Actual Minds, Possible Worlds*. Cambridge: Harvard University Press, 1986.

————. "Vygotsky: A Historical and Conceptual Perspective." In *Culture, Communication, and Cognition: Vygotskian Perspectives*, edited by J. V. Wertsch, pp. 21–34. Cambridge: Cambridge University Press, 1985.

Cassirer, E. *The Philosophy of Symbolic Forms*. Vol. 2, *Mythical Thought*. New Haven: Yale University Press, 1955.

Davydov, V. V., and V. P. Zinchenko. "The Principle of Development in Psychology." *Soviet Psychology* 20 (1981): 22–46.

Donald, M. *Origins of the Modern Mind: Three Stages in the Evolution of Culture and Cognition*. Cambridge: Harvard University Press, 1991.

Edwards, A. D., and V. J. Furlong. *The Language of Teaching*. London: Heinemann, 1978.

Elias, N. *The Civilizing Process: The Development of Manners*. New York: Urizen Books, 1978.

Geertz, C. *Local Knowledge: Further Essays in Interpretive Anthropology*. New York: Basic Books, 1983.

Goodnow, J. J. "Direction of Post-Vygotskian Research." In *Contexts for Learning: Sociocultural Dynamics in Children's Development*, edited by E. A. Forman, N. Minick, and C. A. Stone, pp. 369–81. Oxford: Oxford University Press, 1993.

Habermas, J. *The Theory of Communicative Action*. London: Heinemann, 1984.

Halliday, M. A. K. *Language as Social Semiotic*. London: Edward Arnold, 1978.

Havelock, E. A. *The Muse Learns to Write: Reflections on Orality and Literacy from Antiquity to the Present*. New Haven: Yale University Press, 1986.

Hutchins, E. "The Social Organization of Distributed Cognition." In *Perspectives on Socially Shared Cognition*, edited by L. B. Resnick, J. M. Levine, and S. D. Teasley, pp. 283–307. Washington, D.C.: American Psychological Association, 1991.

Lave, J., and E. Wenger. *Situated Learning: Legitimate Peripheral Participation*. Cambridge: Cambridge University Press, 1991.

Lemke, J. "Genres, Semantics, and Classroom Education." *Linguistics and Education* 1 (1988): 81–99.

Piaget, J. *La psychologie de l'intelligence*. Paris: Colin, 1961.

————. *The Grasp of Consciousness: Action and Concept in the Young Child*. Cambridge: Harvard University Press, 1976.

Raeithel, A. "On the Ethnography of Cooperative Work." In *Communication and Cognition at Work*, edited by Y. Engeström and D. Middleton. London: SAGE, forthcoming.

————. "Production of Reality and Construction of Possibilities: Activity Theoretical Answers to the Challenge of Radical Constructivism." *Activity Theory* 5/6 (1990): 30–43.

Sahlins, M. *Culture and Practical Reason*. Chicago: University of Chicago Press, 1976.

Seeger, F., and H. Steinbring. "The Myth of Mathematics." In *Constructing Mathematical Knowledge: Epistemology and Mathematics Education*, edited by P. Ernest, pp. 151–69. London: Falmer Press, 1994.

Selz, O. *Die Gesetze der produktiven und reproduktiven*. Geistestätigkeit Bonn: Cohen, 1924.

Vygotsky, L. S. *Mind in Society: The Development of Higher Psychological Processes*. Cambridge: Harvard University Press, 1978.

———. "The Genesis of Higher Mental Functions." In *The Concept of Activity in Soviet Psychology*, edited by J. V. Wertsch, pp. 744–88. Armonk, N.Y.: M. E. Sharpe, 1981.

———. *Thinking and Speech*. New York: Plenum Press, 1987.

Wertsch, J. V. *Voices of the Mind: A Sociocultural Approach to Mediated Action*. Cambridge: Harvard University Press, 1991.

———. *Vygotsky and the Social Formation of Mind*. Cambridge: Harvard University Press, 1985.

5

From "Stoffdidaktik" to Social Interactionism: An Evolution of Approaches to the Study of Language and Communication in German Mathematics Education Research

Heinz Steinbring
University of Dortmund

T HE way in which language and communication in mathematics classrooms are studied has changed dramatically over the last fifteen years. This chapter illustrates these changes with respect to German didactics.

STOFFDIDAKTIK

The problem of how language, communication, and mathematics teaching are reflected in the German mathematics education scene can be understood only by looking at the predominant awareness of the nature of mathematical knowledge and of mathematics teaching that was held around thirty years ago. This awareness is based on the so-called Stoffdidaktik (didactics of the subject-matter content), according to which the primary task of mathematics education is to prepare mathematics for students; that is, to elementarize, or to simplify methodically, the mathematical knowledge that is pregiven by the scientific mathematics discipline.

Within this approach to mathematics education, the problem of language and communication emerges in the form of an incompatibility between the mathematical technical language and the everyday language of the students (cf. Abele 1988, 1992; Maier 1986; Maier and Bauer 1978; Winter 1978). The main ideas are that the logically neat and precise technical language of the mathematical discipline has to control the natural language and that in the end, the everyday language of the students is subordinate to the technical language. In specific cases of learning and understanding, one has to admit certain "methodic aids." The technical language must not get too formal and too abstract; in some well-defined cases, everyday descriptions might be allowed for technical definitions. But, for instance, difficulties with logic and clear logical reasoning are arguments for the superiority of the exact technical language over the imprecise natural language. If students are allowed to use everyday language, they get into trouble when it comes to using logical connectors and the logical implication, for example, in forming the contrapositive of statements.

Stoffdidaktik is based on a view of language and communication that Lakoff and Johnson describe as the myth of objectivism.

> According to the myth of objectivism, the world is made up of objects; they have well-defined properties, independent of any being who experiences them, and there are fixed relations holding among them at any given point in time. These aspects of the myth of objectivism give rise to a building-block theory of meaning. If the world is made up of well-defined objects, we can give them names in a language. If the objects have well-defined inherent properties, we can have a language with one-place predicates corresponding to each of those properties. And if the objects stand in fixed relations to one another (at least at any given instant), we can have a language with many-place predicates corresponding to each relation.
>
> Assuming that the world is this way and that we have such a language, we can, using the syntax of this language, construct sentences that correspond directly to any situation in the world. The meaning of the whole sentence will be its truth conditions, that is, the conditions under which the sentence can be fitted to some situation. The meaning of the whole sentence will depend entirely on the meanings of its parts and how they fit together. The meanings of its parts will specify what names can pick out what objects and what predicates can pick out what properties and relations. (Lakoff and Johnson 1984, p. 202)

This unreflected understanding of the role of language and communication in the mathematics classroom becomes obvious in requirements like "to give clean and precise definitions," "to make unambiguous denotations," and "to introduce unique and general concepts" in mathematics teaching. (For a critique of the belief that formal precision and abstract unequivocal statements could avoid misunderstandings or even directly improve students' understanding in mathematics, see Jahnke [1984, 1989]).

This approach to mathematics education follows the illusion of a ready-made and hierarchically structured mathematical knowledge and especially the illusion of a precise technical mathematical language as an objective and universal means to describe completely and to reflect all mathematical facts in all details. According to this illusion, learning and teaching mathematics can be organized by developing and optimizing the technical mathematical language and terminology—sometimes with the help of methodical aids and simplifications.

This perception of the role of language and communication in mathematics teaching has been strongly criticized from two positions:

(1) A philosophical and epistemological critique, which pointed to the naïveté of this perception of the nature of mathematical knowledge both as an academic discipline and with respect to educational needs

(2) A sociological and interactionist critique, which stressed the naive comprehension of the processes of teaching and learning mathematics and of the culture of the mathematics classroom

THE PHILOSOPHICAL AND EPISTEMOLOGICAL POSITION

Concerning the first point of criticism, the question of whether it is possible to characterize science as a language has been a major theoretical problem in the history and philosophy of science for centuries. Within Stoffdidaktik this problem was never taken into consideration. But only with discussions about the theoretical nature of mathematical knowledge, especially emphasizing the genetic and the developmental aspects of mathematical knowledge, can the problem of language and mathematics be approached from this fundamental point.

In a historical and epistemological analysis of the debate at the turn of the eighteenth century about the conception of science as a language, Jahnke and Otte developed fundamental insights into the questions of whether mathematics can be regarded as a language and especially whether mathematics, in relation to the problem of teaching, can be reduced to a naive language. First of all, they make some basic remarks on the specific nature of the epistemological status of scientific concepts.

> [M]odern science tends more and more towards an understanding of concepts as no longer substance concepts in the classical sense, but as relation or function concepts.... According to that, concepts are not names or designations of things, but of relationships between things. Accordingly, a concept designating, for instance, one characteristic feature of a thing, does not only refer to this feature, but to the relationships between it and the totality of characteristic features to which it belongs. (Jahnke and Otte 1981, pp. 76–77)

This understanding of mathematical concepts as referring to relations and not to objects or to empirical properties of objects constitutes the basic step toward developing mathematics education into a scientific discipline.

> For didactics, for instance, it is obvious that the didactical problem in its deeper sense, that is in the sense that it is necessary to work on it scientifically, is constituted by the very fact that concepts will reflect relationships, and not things. Analogously, we may state for the problem of the application of science that it will become a real problem only where the relationship between concept and application is no longer quasi self-evident, but where to establish such a relationship requires independent effort. (Jahnke and Otte 1981, pp. 77–78)

This understanding implies that science can no longer be seen as a language.

> Science as a language implies that application of knowledge becomes something almost automatic, raising no problem at all as long as the language developed is sufficiently good, clear, and reasonable. (Jahnke and Otte 1981, p. 83)

Educational problems, especially, have to cope with a specific field of the application of knowledge that cannot be automatized.

> [E]pistemological conceptions prove more and more discipline-specific and are specifically developed on the bases of individual sciences. The application problem and the teaching problem are no longer considered solvable exclusively in the realm of theory; their practical, political, and organizational dimensions are recognized as well. (Jahnke and Otte 1981, p. 83)

The main consequence of this criticism is that mathematics cannot be seen simply as a naive technical language describing pregiven objective facts in an absolute manner; mathematics is not a simple image of the physical world. But mathematics consists of theoretical concepts (defining the existence of ideal objects) that undergo developments and changes, processes of generalization and of reinterpretation; it cannot be conceived of as a technical language or terminology with everlasting objective and authentic descriptions of eternal truths. Mathematics must be seen as a vivid, open language and a means of communication that produces its own metaphors, meanings, and interpretations.

THE SOCIOLOGICAL INTERACTIONIST POSITION

In terms of the sociological interactionist criticism, Stoffdidaktik was never really concerned with everyday mathematics teaching in the sense of the systematic observation and analysis of mathematics lessons or teaching episodes. The interest of Stoffdidaktik in mathematics lessons was restricted to pointing to the discrepancies between the ideal it had set and the everyday teaching-learning processes.

It was only with the research group formed by Bauersfeld that in the last fifteen years, German mathematics-education research started to take everyday mathematics teaching and learning processes seriously into account (cf. Bauersfeld 1978, 1982, 1983, 1988, 1995; Krummheuer 1984, 1988, 1989, 1995; Maier and Voigt 1991, 1994; Voigt 1984a, 1984b, 1985, 1994, 1995). This group developed a social-interactionist approach in the analysis of everyday mathematics teaching processes (see Sierpinska, chapter 2 in this volume). For the first time, everyday mathematics teaching was not regarded primarily as a deficient process but as an autonomous, self-regulatory process with its own mechanisms, structures, dependencies, frames, and patterns of interaction between the students and the teacher in the classroom. Several sensitizing concepts have been developed in this group, which grasped the phenomena of classroom communication: the funnel pattern (Bauersfeld 1978), mathematical framing (Krummheuer 1982), routines and patterns (Voigt 1984b).

From this interaction-theoretic perspective, language and communication gain a totally different role and importance in looking at and evaluating the processes of teaching and learning mathematics. Within classroom communication, everyday language dominates technical mathematical language; the technical descriptions and notations are subordinate to the everyday speech of students and teachers. And further, the autonomous processes of teaching and learning cannot be dictated by the pregiven mathematical content and by the neat and logically precise technical language. This perspective also changes the way in which the relation between objects and symbols is understood. Bauersfeld criticizes the assumption, long upheld in the philosophy of language, that "names/words fit with objects" (Bauersfeld 1995, p. 273). According to him, neither are the objects referred to by the signs and symbols neutral nor are the symbols merely conventionally constituted as names for objects. There is no strict congruence between objects and symbols. But, according to Bauersfeld, the dominance of presuming this congruence is one cause for the failure of the classic methods of teaching:

> The teacher knows and teaches the truth, using language as a representing object and means. Because there is no simple transmission of meaning through language, the students all too often learn to say by routine what they are expected to say in certain defined situations. (Bauersfeld 1995, pp. 275–76)

The social-interactionist perspective views mathematics teaching as an independent culture, which is not determined by the structure of the scientific discipline and the technical language in a direct way. This culture generates a specific school-mathematical language, and very often the mathematical logic of an ideal teaching-learning process becomes replaced by the social logic of interactions (Bauersfeld 1988, p. 38).

Voigt emphasizes the fact that the interactive patterns of teaching and learning alone are no guarantee of successful mathematical learning.

> The risk of defining learning mathematics as "learning how to participate in social practices" … is that a smoothly proceeding classroom discourse may be interpreted as an indication of successful learning. In fact, several studies analyzing discourse processes in detail have concluded that students' attempts to participate in the constitution of traditional patterns of interaction can be an obstacle for learning mathematics.… In usual lessons, the classroom discourse often tends to degenerate into rituals that are constituted step by step. Surprisingly, the teachers are unaware of these regularities in the microprocesses, and they misinterpret the students' participation in the classroom discourse. Therefore, we should resist the temptation to identify learning mathematics with learning how to participate successfully in patterns of interaction. (Voigt 1995, p. 166)

It would seem, therefore, that some main differences between Stoffdidaktik and social interactionism, with regard to language and communication in the mathematics classroom, are that the former concentrates on the syntax of the technical mathematical language, whereas in the latter, "mathematics is a language seen from the perspective of pragmatics not semantics or syntax" (Sierpinska, chapter 2 in this volume). In contrast, the philosophy and epistemology of mathematics take yet another position focusing on the peculiar semantics of mathematical symbols and on the theoretical nature of mathematical knowledge.

AN ILLUSTRATION

In the following section, a short teaching episode from a mathematics lesson (grade 7) is presented. It will be discussed with regard to the three positions mentioned above. The transcript is drawn from a class on the topic of relative frequency.

The Episode

1. *T:* What do we understand by relative frequency, Markus?
2. *S:* Relative frequency is, if you take the number of the cases observed together
3.   with the number of trials, well you throw all into one pot, and stir it up.
4. *S:* Hahaha, enjoy your meal!
5. *T:* Klaus?
6. *S:* Relative frequency means for example often, it is, um, a medium value.
7. *T:* Yes, what ... ? Then it's better to say nothing!
8. *T:* Relative frequency, Frank?
9. *S:* Um, relative frequency does mean, when the trials are divided with the
10.   cases observed, I think, or multiplied.
11. *T:* Markus did already say it quite correct, just the crucial word was missing.
12. *T:* ... But now you do know it, Markus?
13. *S:* Subtract.
14. *S:* Minus it.
15. *T:* That's incredible! Silvia?
16. *S:* In a random experiment, the number of cases observed, when you...
17. *T:* Um, Markus, you should ask how you will write it down?
18. *S:* Oh yes, OK., well, relative frequency is, when you, the number of trials with
19.   the number of cases observed, well, if you then ...
20. *T:* But how you write it down? How do you write it down?
21. *T:* Come on, write it on the blackboard!
22. *S:* [*Shortly writing down*].
23. *T:* Thank you, Ulli.
24. *S:* [*Writing as a fraction.*]
25. *T:* Aha, as a fraction, so, what is it then, what kind of calculation?

26. *S:* Fractional calculation.
27. *T:* Well, a complete sentence, Markus!
28. *S:* Um, well relative frequency is …
29. *T:* Leave it aside.
30. *S:* … The number of, um, the number of cases observed with the number of
31. trials, yes, subtracted in the fractional calculation.
32. *T:* Subtracted in the fractional calculus! Markus, that's incredible!
33. *T:* Ulli, formulate it in a more reasonable way!
34. *S:* If you take that, the number of the, the cases observed, yes, and the number
35. of trials, hem, well, if you then, one as denominator and the other as
36. numerator, hem, I don't know what you mean.
37. *S:* To calculate a fraction!
38. *T:* Either to write as a fraction, or what kind of calculation is it then, if you write a
39. fraction?
40. *S:* Dividing.
41. *T:* Yes, that's what I think too. OK, yes. Nobody knows it any more?!
42. *S:* To reduce to the common denominator.
43. *T:* Well, relative frequency is, and Markus has said it correctly, the number of
44. cases divided by .. ?
45. *S:* .. The number of cases observed.
46. *T:* No, the cases observed divided by ? Aha! Or, if you formulate it as Ulli
47. said, the cases observed as numerator and the number of trials as ?
48. *S:* Denominator.

This episode can be structured roughly in three phases. The opening question of the teacher can be seen as the first phase (1). In the second phase (2–16), the teacher tries to elicit the correct verbal description of relative frequency. The proposals made by Markus, Klaus, Frank, and others furnish all but one missing element. In the third phase (17–48), the teacher has the students write down the definition of relative frequency on the chalkboard. The teacher is especially keen on having it written as a fraction. This classroom interaction leads in the end to an accepted solution.

At the beginning, the teacher tries to recall the concept of relative frequency, starting with the question, "What do we understand by relative frequency, Markus?" (1). Already the way of formulating "what do *we* understand …" indicates some accepted, conventionalized form of denoting this concept. The first student, Markus, brings together nearly all the necessary elements—"the number of cases observed" and "the number of trials"—and he points out humorously that these elements have to be combined.

Then Klaus proceeds in another direction, saying, "Relative frequency means … often … a medium value." (6). The teacher rejects his proposal, and Frank goes on with the old idea, trying some possibilities for combining the elements already identified (9,10). This response is strongly approved by the teacher (11), which clarifies the accepted frame for the students to handle the question. They are expected to find the needed pieces of a puzzle. Two pieces, "the number of cases observed" and "the number of trials," have already been found. Another decisive piece, the mathematical combination of these two, is still missing. The spontaneous suggestions "subtract" and "minus it" provoke a forceful rejection by the teacher (15), which leads to an alternative approach that asks for the way this concept should be written down.

The third phase introduces the context of fraction and fractional calculus into the discussion. Contrary to her expectations, the key words *fraction* and *fractional calculation* (24, 26) evoke a whole variety of students' contributions referring to many fractional operations (31, 35, 36, 37, 42). The teacher simply intended to point to the link between a fraction and division, a seemingly simple transfer that could be made only toward the end of this phase in a funnel-like pattern (Bauersfeld 1978).

When the last piece of the puzzle is found, combining all three in the right way remains (43–48). The first offer made by the teacher (43, 44) does not use the detailed vocabulary, and, consequently, the answer of a student does not fit (45). At last the correct formulation is initiated by the repeated question, "No, the cases observed divided by … ? Aha!" (46). Finally, relative frequency is also formulated in terms of fractional calculus.

Analysis

How could this episode be interpreted from the points of view of Stoffdidaktik, social-interactionist theory, and epistemology?

An analysis from the point of view of Stoffdidaktik

Stoffdidaktik certainly would strongly criticize the totally unclear and imprecise words, descriptions, and concepts the students are using. Also, the teacher in this episode would be judged guilty of a bad use of language, that is, the natural language used here, and of the lack of any technical language and conceptual description. Using such language, according to Stoffdidaktik, would certainly help avoid and clarify this disordered situation of classroom communication.

The mathematical concept looked for in this episode—the concept of relative frequency—seems not to be introduced clearly and correctly in a way that the students are able to name and use without any problems and to see immediately the links to the fraction concept. The teacher in this situation does not give adequate hints, help, or clear questions, which would allow the students to find a mathematically satisfying answer; she herself is using rather unclear and sloppy natural language.

Thus, for Stoffdidaktik this episode is an example of rather bad mathematics teaching that shows clearly the deficits and differences with respect to an ideal model of teaching and learning that uses precise notations and technical definitions. Only the exactness of the mathematical language could avoid such a disastrous teaching situation.

An analysis from the point of view of social interactionism

From the perspective of the social-interactionist theory, this episode would be taken seriously as it is. It represents usual, everyday mathematics teaching-learning process. One can identify different communicative patterns and routines in the course of interaction, for instance, the funnel pattern as the teacher narrows down her questions to force the correct answer when searching for the only missing word, *divided* (38, 39, 40). Another pattern that can be identified is related to the different frames that the teacher and the students hold with the introduction of the fraction concept as a kind of reference for the concept of relative frequency.

The communicative process between students and teacher in this episode emerges as an autonomous self-regulating process, which shows how the different contributions depend on one another. The teacher cannot simply escape this communicative involvement by using a seemingly correct technical mathematical language or by presenting correct technical mathematical definitions from the outside. This dependence between different contributions is obvious, and it shows that many students are very successful here in participating in the patterns of interaction. This could best be exemplified in our episode by the following unsuccessful interactive pattern. At an early stage in this episode, the expected answer could have been given. Frank answers the teacher's question by presenting more clearly some possibilities for the mathematical operations that are looked for, "divided" and "multiplied," whereas Markus seems to point out only that some mathematical operation is still missing.

But the teacher does not follow exactly the pattern of interactive communication that students would expect here and that they usually get as reaction from the teacher. She does not—implicitly or explicitly—signal that one of the possibilities offered by Frank—"divided" or "multiplied"—could be the correct answer. She rejects his contribution totally by referring back to Markus's contribution. Perhaps in this way the teacher leads the students astray and thus produces some of the problems that can be observed in the long process of finding the expected answer.

An epistemological analysis

An epistemological perspective on this teaching episode would also consider this communicative process independent and autonomous, an interaction that constitutes its proper knowledge, concepts, and meaning. Its main question would be, however, What is the epistemological status of the thus constituted school mathematical knowledge?

One instrument of analysis is the epistemological triangle (cf. Steinbring 1988, 1989, 1991a, 1991b, 1994a, 1994b, 1994c, 1997), which tries to identify different kinds of relationships that are constructed during the classroom discourse between signs or symbols and contexts of reference. A major epistemological problem during this episode is how to endow the concept of relative frequency with meaning. In a more concrete sense, the students are asked to explain the "name" of relative frequency with the help of known notations. In a first approach, some of the notations involved in the definition of relative frequency are collected: "the number of the cases observed" and "the number of trials." These two have to be combined mathematically. Behind this attempt is the official definition already dealt with in an earlier lesson: Relative frequency is the number of cases observed divided by the number of trials. This verbal description is often visualized as a fraction in the following manner:

$$\text{relative frequency} = \frac{\text{number of cases observed}}{\text{number of trials}}$$

According to the epistemological triangle of concept, sign or symbol, and context of reference (fig. 5.1), the interaction observed in the course of this episode concentrates on establishing relationships between sign or symbol and context of reference (fig. 5.2). The students are expected to give an explication of the notation (the name) of relative frequency with reference to the elements {number of cases observed, number of trials, divided} together with the correct combination or structure of these elements. Because in the beginning, the

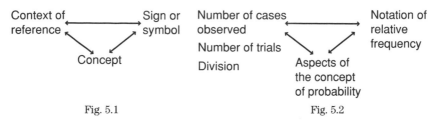

Fig. 5.1   Fig. 5.2

students are unable to find all the correct elements and their combination, the teacher suggests that they write down the formula for relative frequency. A student completes her suggestion by saying that it should be written as a fraction. This response is reinforced by the teacher.

In this interaction, mathematical concepts are constituted as some kind of "names" for objects, and these names have to be defined by enumerating their properties. In the communicative process, this kind of concept definition degenerates into a guessing game, devoid of real mathematical meaning, which takes into account only the syntactical structure of the three symbolic elements written in the manner of a fraction.

The signals given by the teacher in the course of this episode intend to orient the students toward the conventionally agreed-on definition of the concept of relative frequency. This intention is understood very quickly by most students, and the rejection of Klaus's proposal, perhaps leading away from this orientation, reinforces the conveying of this intention. The more the accepted answers are restricted to the syntactical and conventional aspects, the more the students are deprived of some epistemological means of justification. They are able to make only proposals to be approved or rejected by the teacher. The epistemological helplessness of the students becomes obvious when they understand the signal "fraction" too extensively instead of simply reading off the operation of division from the fraction bar.

The epistemological analysis of this episode suggests that the mathematical knowledge that is constituted in the interaction between teacher and students is of a specific empirical type. But the nature of the concept of relative frequency is theoretical, like all mathematical knowledge. The communication and interaction taking place in this episode cannot and do not cope with an open, developmental constitution of mathematical concepts as relational structures. The epistemological problem of the interactively constituted mathematical knowledge in the course of this episode is that a nontheoretical understanding of mathematical knowledge is established, which hinders real mathematical understanding and blocks the generation of meaning.

From the epistemological perspective, one might notice also that in this episode the students are expected to introduce a mathematical relation (a theoretical structure) between the two already identified elements—"number of cases observed" and "number of trials"—namely, the relation "divided." But this theoretical structural element "divided" is simply taken here as a word, as a syntactical element in the verbal combination to explain the notation of relative frequency. "Divided" is not interactively discussed here as a mathematical relation, for instance, as expressing a relation between the whole and a part or as a specific "proportion," a "relative part" of some "wholeness."

The meaning of theoretical mathematical knowledge can be constituted in the course of interactive processes only if the meaning of mathematical knowledge is not confined to the actual interactive social practice in which students and teacher are involved and if it is accepted that there is some kind of epistemological meaning of mathematical knowledge "outside" the actual social practice. Thus, the epistemological perspective on classroom discourse means looking for the specific semantics of mathematical knowledge that is constituted in the interactive process and in what way the constituted meaning matches a rich semantical structure of theoretical mathematical knowledge.

The social-interactionist, as well as the epistemological, perspective disagrees with the naive position of Stoffdidaktik according to which a ready-made hierarchical structure of mathematical knowledge together with an exact technical language and terminology is the fundamental basis for successful teaching-learning processes. Mathematical meaning is not constructed with the help of formal structures and precise formal definitions. Further, social interactionism denies that any meaning of mathematical knowledge could exist outside a social practice; the constitution of meaning always depends on participation in a social practice. The epistemology of mathematics claims that theoretical meaning also exceeds the actual social practice and thus "exists" outside the borders of the already known social habits and practices.

FINAL REMARKS

In summary, the analysis of the three educational positions toward language and communication in the mathematics classroom, as developed in German mathematics educational research, shows a differentiation in the ways that the social processes of teaching and learning and the body of mathematical knowledge and concepts depend on each other (fig. 5.3).

Fig. 5.3

Stoffdidaktik considers "the body of mathematical knowledge and concepts" as predominant and regards language and communication in the mathematics classroom as dependent on, and determined by, the given body of mathematical knowledge. Social interactionism emphasizes the independence of "the social processes of teaching and learning" and their autonomy in the constitution of meaning. The epistemological approach holds the view that theoretical mathematical meaning cannot be limited to the actual social practice but has to take into account conceptual relations in the mathematical knowledge not yet negotiated in the social practice.

These three positions take different standpoints on the nature and the role of language for teaching and learning mathematics. Stoffdidaktik sees language, especially mathematical language and mathematical signs, symbols, and notations, as some kind of codified signs for objectively existing mathematical facts and objects. For the social interactionist, language is the central means in a social practice to bring about any meaning and to communicate meaning and knowledge. As Sierpinska (chapter 2 in this volume) referring to Bruner (1985) puts it, "Mathematics is ... a discourse.... It is a way of seeing the world, and thinking about it ..., a universe established through communication, whereby people commit themselves to certain conventions, build shared understandings of contexts, and develop conventional means for establishing and retrieving presuppositions."

In the epistemological approach, mathematics is also a discourse but it is not subject to the social practice alone. Of course, there are social conventions and socially shared understandings, but there are also epistemological constraints on the interpretation and meaning of theoretical mathematical knowledge that remain principally unchanged in its epistemological core over different social practices. Religious discourses constituted in different social practices might become very different and disconnected ways of seeing the world. But a mathematical discourse in some fundamental epistemological relationships remains invariant over different social practices in its intended generalization.

Or to say it from the other side: If the meaning of mathematical language is restricted too closely and bound too directly to a specific social practice, there is a risk of destroying theoretical mathematical meaning by a reduction and a hypostasis of mathematical relations instead of inducing an enrichment of meaning by the interactive construction of new and more general relations.

REFERENCES

Abele, A. "Kommunikationsprozesse im Mathematikunterricht" (Communication process in mathematical education). *Mathematische Unterrichtspraxis* 9 (1988): 23–30.

———. "Schülersprache—Lehrersprache: Zwei Fallstudien aus dem Mathematikunterricht" (Student's language–teacher's language: Two case studies from a mathematical lesson). *Mathematische Unterrichtspraxis* 13 (1992): 1–8.

Bauersfeld, H. "Kommunikationsmuster im Mathematikunterricht—Eine Analyse am Beispiel der Handlungsverengung durch Antworterwartung" (Patterns of communication in mathematics teaching: An analysis with the example of narrowed fields of activity by expected outcomes). In *Fallstudien und Analysen zum Mathematikunterricht*, edited by H. Bauersfeld et al., pp. 158–70. Hannover: Schrödel, 1978.

———. "Analysen zur Kommunikation im Mathematikunterricht" (Analyses of communication in mathematics classroom). In *Analysen zum Unterrichtshandeln*, edited by H. Bauersfeld et al., pp. 1–40. Cologne: Aulis, 1982.

———. "Subjektive Erfahrungsbereiche als Grundlage einer Interaktionstheorie des Mathematiklernens und lehrens." In *Lernen und Lehren von Mathematik*, edited by H. Bauersfeld et al., pp. 1–56. Cologne: Aulis, 1983.

———. "Interaction, Construction, and Knowledge: Alternative Perspectives for Mathematics Education." In *Effective Mathematics Teaching*, edited by D. A. Grouws, T. J. Cooney, and D. Jones, pp. 2–46. Reston, Va.: National Council of Teachers of Mathematics and Hillsdale, N.J.: Lawrence Erlbaum Associates, 1988.

———. "'Language Games' in the Mathematics Classroom: Their Function and Their Effects." In *The Emergence of Mathematical Meaning: Interaction in Classroom Cultures*, vol. 2, edited by P. Cobb and H. Bauersfeld, pp. 271–91. Hillsdale, N.J.: Lawrence Erlbaum Associates, 1995.

Bruner, J. S. "The Role of Interaction Formats in Language Acquisition." In *Language and Social Situations*, edited by J. P. Forgas, pp. 31–46. New York: Springer-Verlag, 1985.

Jahnke, H. N. "Anschauung und Begründung in der Schulmathematik." *Beiträge zum Mathematikunterricht* (1984): 32–41.

————. *Abstrakte Anschauung, Geschichte und didaktische Bedeutung.* Occasional Paper Nr. 115. Bielefeld: Inst. für Didaktik der Mathematik, 1989.

Jahnke, H. N., and M. Otte. "On 'Science as a Language.'" In *Epistemological and Social Problems of the Sciences in the Early Nineteenth Century*, edited by H. N. Jahnke and M. Otte, pp. 75–89. Dordrecht, Netherlands: Reidel, 1981.

Krummheuer, G. "Rahmenanalyse zum Unterricht einer achten Klasse über 'Termumformungen.'" In *Analysen zum Unterrichtshandeln*, edited by H. Bauersfeld et al., pp. 41–103. Cologne: Aulis, 1982.

————. "Zur unterrichtsmethodischen Diskussion von Rahmungsprozessen" (Methodical dimension of frame processes). *Journal für Mathematik Didaktik* 5 (1984): 285–306.

————. "Verständigungsprobleme im Mathematikunterricht" (Communication problems in mathematics teaching). *Der Mathematikunterricht* 34 (1988): 55–60.

————. "Die Veranschaulichung als 'formatierte' Argumentation im Mathematikunterricht" (Illustration as formated argumentation in mathematics teaching). *Mathematica Didactica* 12 (1989): 225–43.

———— "The Ethnography of Argumentation." In *The Emergence of Mathematical Meaning: Interaction in Classroom Cultures*, edited by P. Cobb and H. Bauersfeld, pp. 229–69. Hillsdale, N.J.: Lawrence Erlbaum Associates, 1995.

Lakoff, G., and M. Johnson. *Metaphors We Live By.* Chicago: University of Chicago Press, 1984.

Maier, H. "Empirische Arbeiten zum Problemfeld Sprache im Mathematikunterricht" (Empirical investigations of the language in the mathematics classroom). *Zentralblatt für Didaktik der Mathematik* 18 (1986): 137–47.

Maier, H, and L. Bauer. "Zum Problem der Fachsprache im Mathematikunterricht" (The problem of technical language in mathematics teaching). In *Conference: Learning Difficulties in Mathematics Teaching, Bielefeld, Germany*, 21–23 November 1978, Schriftenreihe des IDM, edited by H. Bauersfeld, M. Otte, and H. G. Steiner, pp. 137–59. Bielefeld: Inst. für Didaktik der Mathematik, 1978.

Maier, H., and J. Voigt, eds. *Interpretative Unterrichtsforschung.* Cologne: Aulis, 1991.

————. *Verstehen und Verständigung im Mathematikunterricht: Arbeiten zur interpretativen Unterrichtsforschung.* Cologne: Aulis, 1994.

Steinbring, H. "Nature du savoir mathématique dans la pratique de l'enseignant." In *Actes du Premier Colloque Franco-Allemand de Didactique des Mathématiques et de l'Informatique*, edited by C. Laborde, pp. 307–16. Grenoble: La Pensée Sauvage, 1988.

————. "Routine and Meaning in the Mathematics Classroom." *For the Learning of Mathematics* 9 (1989): 24–33.

————. "The Concept of Chance in Everyday Teaching: Aspects of a Social Epistemology of Mathematical Knowledge." *Educational Studies in Mathematics* 22 (1991a): 503–22.

————. "Mathematics in Teaching Processes: The Disparity between Teacher and Student Knowledge." *Recherches en Didactique des Mathématiques* 11 (1991b): 65–107.

————. "Symbole, Referenzkontexte und die Konstruktion mathematischer Bedeutung—am Beispiel der negativen Zahlen im Unterricht." *Journal für Mathematik–Didaktik* 15 (1994a): 277–309.

————. "Dialogue between Theory and Practice in Mathematics Education." In *Didactics of Mathematics as a Scientific Discipline*, edited by R. R. Biehler, R. W. Scholz, R. Sträßer, and B. Winkelman, pp. 89–102. Dordrecht, Netherlands: Kluwer Academic Publishers, 1994b.

————. "Die Verwendung strukturierter Diagramme im Arithmetikunterricht der Grundschule—Zum Unterschied zwischen empirischer und theoretischer Mehrdeutigkeit mathematischer Zeichen." *Mathematische Unterrichtspraxis* 15 (1994c): 7–19.

————. "Epistemological Investigation of Classroom Interaction in Elememtary Mathematics Teaching." *Educational Studies in Mathematics* 32 (1997): 49–92.

Voigt, J. "Die Kluft zwischen didaktischen Maximen und ihrer Verwirklichung im Mathematikunterricht. Dargestellt an einer Szene aus dem alltäglichen Mathematikunterricht" (The gap between didactic conceptions and the practice of mathematical instruction. Illustrated by an example from regular mathematics instruction). *Journal für Mathematik-Didaktik* 5 (1984a): 265–83.

————. *Interaktionsmuster und Routinen im Mathematikunterricht— Theoretische Grundlagen und mikroethnographische Falluntersuchungen.* (Interaction patterns and routine in instructing. theoretical basics and microethnographical case studies). Weinheim: Beltz, 1984b.

————. "Patterns and Routines in Classroom Interaction." *Recherches en Didactique des Mathématiques* 6 (1985): 69–118.

————. "Negotiation of Mathematical Meaning and Learning Mathematics." *Educational Studies in Mathematics* 26 (1994): 275–98.

————. "Thematic Patterns of Interaction and Sociomathematical Norms." In *The Emergence of Mathematical Meaning: Interaction in Classroom Cultures*, edited by P. Cobb and H. Bauersfeld, pp. 163–201. Hillsdale, N.J.: Lawrence Erlbaum Associates, 1995.

Winter, H. "Umgangssprache–Fachsprache im Mathematikunterricht" (Colloquial language–technical language in mathematics teaching). In *Conference: Learning Difficulties in Mathematics Teaching, Bielefeld, Germany*, 21–23 Nov 1978, Schriftenreihe des IDM, edited by H. Bauersfeld, M. Otte, and H. G. Steiner, pp. 5–56. Bielefeld: Inst. für Didaktik der Mathematik, 1978.

6

Examing the Linguistic Mediation of Pedogogic Interactions in Mathematics

Clive Kanes

Griffith University, Brisbane

W HAT is the nature of communication in mathematics teaching? How does communication relate to the workings of language? How is mathematical knowledge related to language? In this chapter, the philosophy of language suggested by Ludwig Wittgenstein (1889–1951) in his work *Philosophical Investigations* (1991) is used to prompt discussion related to these questions. As a starting point, consider the following observation (fig. 6.1) made by Wittgenstein in his *Remarks on the Foundations of Mathematics* (1967, §14):

You only need to look at the figure

to see that 2 + 2 are 4. —Then I only need to look at the figure

to see that 2 + 2 + 2 are 4.

Fig. 6.1

The imaginary conversation described in this remark illustrates an all too familiar situation. Even after the teacher devises an apparently flawless scheme for instructional clarity, students sometimes still do not appear to grasp the ideas placed before them. Moreover, tools commonly used for mathematics instruction, such as diagrams, graphs, and concrete aids, seem to acquire a life of their own, once introduced into the pedagogic interaction. Teachers intend them to be used in one way, yet their presence in the classroom leaves open the possibility that students may use them in ways that can run at a tangent to, and possibly undermine, the purposes of instruction. These observations lead to questions about how communication works and, more to the point in light of the comments above, how it happens that communication sometimes fails. Now, in addressing these questions, a conventional response would be to argue that misunderstanding a received message is based on applying the wrong process of decoding the speaker's language. This line entails the understanding that language operates like a code in which messages are first formed by the speaker, then encoded within an appropriate medium and transmitted. Encryptions, in turn, are received and decoded by the interlocutor, which enables the message to be read. On this view, misunderstandings are merely errors induced by the breakdown either of the correct technical procedures for encryption or of the means whereby the coded message is transmitted. Although Wittgenstein recognizes that in some respects language is codelike, he maintains that this view of language is far from adequate and therefore cannot act as a basis for understanding the circumstances in which complex communication succeeds and sometimes fails.

WITTGENSTEIN'S VIEWS ON LANGUAGE AND COMMUNICATION

Central to Wittgenstein's alternative account is the belief that language has a double-sided character. On one side, straightforwardly enough, it can be like a reporting system that passes us information about events that occur at arm's length; yet, on the other side, language can interact with the message itself. On this view, the objects of language (information, thoughts, ideas, points of view) become entangled in the processes of their articulation and communication and thus become a part-product of language itself. Therefore, to understand

fully a thought or our ideas on a particular subject, we need to understand how the language we use comes to be involved in the process of
rendering our thinking. Likewise, when we want to understand more
about how to communicate a thought, we must also understand more
about how language works to form both our thinking and our efforts to
communicate with others.

This insight provides the basis for Wittgenstein's explanation of how
communication works and, as important from the viewpoint of the
teacher, when and how communication fails. For Wittgenstein, language has an impact on how meaning is produced in two ways. First, it
does so by imprinting its own logical form or *grammar* on the concept-
world it generates. Thus, philosophical investigations, interpreted as
investigations into the concepts that organize our thinking, become primarily grammatical investigation, that is, investigations concerning the
implications for meaning of the logical content of language and its operation. Concerning teaching, this suggests that whenever the purpose is
to consider how meaning is generated, such as in typical pedagogic
interactions, attention could and perhaps should be directed to the
function and structure of language as the medium of the exchange as
well as to the content—the mathematical concepts. These last, in most
practical situations, are normally supposed to lie outside the linguistic
domain. In his discussion of grammar, it should be emphasized,
Wittgenstein is not, of course, reflecting on syntactical structures of
sentences. Instead, he is seeking to identify the logical structures that
language uses to make communication possible.

The second way, according to Wittgenstein, that language influences
meaning relates to the particular context of language use rather than to
the logic or structure of language. For Wittgenstein, meaningful language
is thought to perform an action with respect to the purposes the speaker
intends. To grasp meaning correctly, therefore, the listener or reader
must participate in the *language-game* within which the activities of the
language user are realized. For instance, in figure 6.1, the failure to understand the correct rules of the counting game (in the sense intended in the
communicative interaction) leads the interlocutor into the apparent mistake of double counting a number of crosses in the diagram. This example illustrates that even in relatively straightforward instances, meaning
does not lie inherently within the language system as such, but within the
particular circumstances whereby given utterances are actually used.
Nothing in the diagram in figure 6.1 enables the viewer simply to merely
look and see the correct procedure to be followed. For the interlocutor,

the knowledge of what meaning to take from the encounter is supplementary to the formal components of the engagement.

In summary then, language is double-sided in its operations. Double-sidedness contributes to the success of language as a means for communication and also explains ways in which this communication becomes faulty. Roughly speaking, these failures occur (1) when the speaker or user of language is misled by semantic distortions introduced by the action of grammatical constructions required to make communication possible or (2) when communication is predicated on a language-game (e.g., "decoding" as above) in which, in fact, parties to the episode are not jointly engaged. An aim of this chapter is to illustrate these points and thereby make them available for further discussion and analysis. Given that Wittgensteinian analyses have generally occupied a peripheral position within the arena of mathematics education research (Bloor 1976, 1983; Confrey 1981; Hamlyn 1989; Kanes 1993; Watson 1988, 1989; Bauersfeld 1995), much of the work of this chapter must be directed toward introducing the distinctive approach adopted by these perspectives. Care has been taken to locate these developments within a framework relating to alternative traditions currently more influential within the field of mathematics education research. Considering the scale of these tasks, however, the study of the curriculum implications of these perspectives occupies a less prominent place in this work than might otherwise have been desired. Such analyses would be suitable for a more extended exposition than is possible here.

The structure of the chapter is as follows. In the next section, the aim is to provide a theoretical background and rationale for the so-called *linguistic turn* against which the Wittgensteinian philosophy, sponsored in this work, may be understood. Other sections in the chapter introduce the Wittgensteinian framework, use this to analyze the communicational component of an actual mathematics pedagogic interaction, and draw final conclusions from this study that relate to communication within the domain of mathematics teaching.

TEACHING AND LEARNING IN MATHEMATICS: FROM THE EPISTEMO-LOGICAL TO THE LINGUISTIC DOMAIN

When considering language in the circumstances of teaching and learning, researchers have shown a marked tendency to consider only the information-carrying side of the operation of language (see, i.e., Resnick [1983]; Pimm [1987]; Ellerton and Clements [1991]). In Wittgensteinian terms, such analyses are one-sided in that they emphasize too strongly the technical requirements needed for achieving accurate meanings and downplay the influence of language itself in forming the message communicated. In Resnick's view, for instance, "target" knowledge is to be encoded in instructional representations that when properly designed enable the learner to grasp the teacher's conceptual intention. Transparency with respect to the medium is the condition for optimal performance of this type of linguistic technology. One well-rehearsed line of criticism of this model has been based on challenges to the validity of the epistemological assumptions on which it is grounded. Radical constructivism, for instance, contrasts with the so-called naive constructivism of Resnick and others by denying claims that mathematical concepts belong to a real, albeit abstract, domain (Cobb 1988, 1992; Ernest 1991; von Glasersfeld 1991). Thus, not only is it not clear what the object of a mental representation of a concept could logically be, but, even more seriously, it is uncertain how the learner's mind is able to identify the target concept encoded within pedagogically tailored representations (Cobb 1992).

A second line of criticism of this overtly technical view of communication attempts to move beyond the epistemological questions characteristic of the constructivist paradigm (see, i.e., Kanes [1993, 1996a, 1996b, 1997a, 1997b, in press]). Within this general approach, as noted above, various linguistic frameworks have been developed, each involving, in various formulations, its own version of the concept of the double-sided character of language. As background to the direction taken in this chapter, a brief survey of the lines of development taken by these approaches follows. Three separate traditions have appeared. The first takes, as a starting point, Saussure's theory of structural linguistics and includes the work of the contemporary poststructuralist school. The second relates to the concept of "voice" in the work of Bakhtin (1986) and to what Holquist (1990) and Wertsch (1991) have called *dialogism*. The third follows the path outlined in this chapter and is influenced by a branch of linguistic philosophy known as speech act theory. This last theoretical framework was first developed by Wittgenstein and later taken up by the British philosophers Austin (1980), Strawson (1971), and Ryle (1971).

Saussure's approach (1983) focuses on the microfoundations of meaning and rests on the view that language operates differentially (that is, as a structure in which words are defined with respect to how their meanings differ) and works by forming a representation of intended concepts underlying communicative processes. Post-structural linguists reject both aspects of this project. The works of the French theorists Barthes (1968), Kristeva (1986, 1989), Derrida (1976), and Lacan (1977), for instance, are most notable studies in this tradition.

Common among these authors is a belief in the radical nonclosure of texts as a site for meaning making. Foucault's work (1972) also developed in part from that of Saussure, although his preoccupations were with the macrofoundations rather than the microfoundations of meaning. Starting with the Saussureian concept of discourse (a signifying practice referring to contents or representations), he extended the notion to describe a practice that "systematically forms the objects of which it speaks" (Foucault 1972, p. 49). Foucault argued that concepts do not belong to an external reality but to the social organization of language in which discursive relations are practiced and have life. It is notable that discussions concerning the use of language in pedagogic situations have generally not engaged the theoretical issues raised by Saussureian and post-Saussureian linguistics. Although this work clearly cannot be achieved here, it is of importance to note that important contributions to understanding pedagogic interactions could be generated through approaches of this kind (e.g., Walkerdine [1988, 1994]). Although meaning may be unstable, meanings generated in discourse can carry profound consequences for the lives of individuals and for the collectives to which they belong.

The second tradition of linguistic analysis to warrant attention is that of the Russian school. This approach has been built on the understanding of language as a tool in the social formation of the mind as represented paradigmatically in the work of Vygotsky (1978, 1987) and Bakhtin (1986). In contrast to the Saussureian tradition, speech (parole) rather than the formal structures of language (langue) is the focal point for research. In this view, language, even when internalized and represented as thought, is always a "turning to someone else" (Wertsch 1991) and is therefore irrepressibly multivocal. At the root of all higher mental functioning is language, and the root of all language is dialogue—the primordial basis of communicative possibility. Although it is not the aim of this chapter to explore these themes (however, see Seeger, chapter 4 in this volume), pedagogic interactions in mathematics can, and also should, be studied with respect to these perspectives.

The third tradition of linguistic analysis follows the path first established by Wittgenstein and is highlighted in the current work. As previously discussed, this approach views speech as first and foremost an action. Although issues of formal linguistic structure, or even the dialogic flavor of language, are not strictly pertinent to this perspective, much could be learned through a discussion concerning the interrelationship among these traditions (for contrasting views on the relationship between the Wittgensteinian approach and post-Saussureianism see Norris [1983]; Staten [1986]; Ellis [1989]).

In the following section, a general introduction to Wittgensteinian analysis is given.

A WITTGENSTEINIAN FRAMEWORK FOR THE STUDY OF MATHEMATICAL LANGUAGE

For Wittgenstein, mathematical realities are not mirrored in language, they are a grammatical construction of language. This means that language cannot be viewed merely as a technical apparatus whereby facts (specifically mathematical) about the world are rendered communicable. Language is also a constructive means to understanding and knowing, and its grammars, structures, and conventions help to generate and sustain the categories of thought that organize and constitute our thinking. To make these ideas clearer, particularly as they relate to Wittgenstein's work, Wittgenstein's earlier and later periods of thought concerning language and its relationship to reality are contrasted.

In his first published work, the *Tractatus Logico-Philosphicus* (1961), Wittgenstein advanced the thesis that propositions offer us "pictures" of states of affairs that hold in reality. They functioned, he argued, by simply pointing out the facts to which they refer. For instance, one might say that the mathematical proposition $2 + 2 = 4$ is valid because it is a truthful representation of the logical form inherent within, for instance, figure 6.2.

In this account, language operates by forming pictures of reality that faithfully reproduce its logical relationships. Within mathematics education, particularly in the formation of effective teaching methods, adherence among theorists to this doctrine has been, until recently, quite common. For instance, Dienes's idea of the embodiment of a mathematical concept within instructional materials relies on the notion of the logical isomorphism between concrete and abstracted

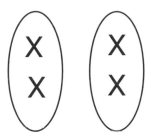

Fig. 6.2

modes of mathematical representation (Dienes 1963). Likewise, Resnick's notion of transparent pedagogical representations (Resnick 1983) refers to similar ideas.

In his later work, however, Wittgenstein abandoned this representational view of language. He argued that it offered, at best, a restrictive view of how language operates and that it leads to serious errors when such questions as the nature of knowledge and the mental processes said to underpin psychological verbs, such as *to understand, to intend, to mean, to think,* and *to know,* are considered. Words, for Wittgenstein, are not names corresponding to objects as much as tools capable of performing tasks.

> Think of the tools in a tool-box: There is a hammer, pliers, a saw, a screwdriver, a rule, a glue-pot, glue, nails and screws. —The functions of words are as diverse as the functions of these objects. (And in both cases there are similarities.)
>
> Of course, what confuses us is the uniform appearance of words when we hear them spoken or meet them in script and print. For their application is not presented to us so clearly. Especially when we are doing philosophy! (Wittgenstein 1991, §11)

Crucial then, for Wittgenstein, is the full investigation of all aspects of language use, including what, when, how, and why words are used by people in circumstances in which various forms of communication are being attempted. In such analyses, Wittgenstein places the specific context of an utterance in a central position. Language, he finds, does not in general work like a cipher. It does not engage abstract benchmarks to fashion the terms of its carriage of truth. As universal standards are logically separated from the particular circumstances, or *forms of life,* that confer specificity to the linguistic acts, Wittgenstein considers them to be useless in any explanation of how meanings are established by words. In contrast, Wittgenstein's aim is to "bring into prominence that the speaking of a language is part of an activity, or form of life" (1991, §23). This insight represents a major departure from earlier theories of language because it emphasizes that speaking a language is part of a life-centered activity or language-game, rather than simply an extended code or representation of reality (1991, §13).

Three senses are distinguished for the language-game concept in his work. The first fixes attention on activities in which language is deployed very directly and purposefully, for instance, in the communications between a builder and assistant. When the builder wants a block, pillar, slab, or beam, he or she calls out "block," "pillar," "slab," "beam." Such a simplified (and highly artificial) language activity Wittgenstein refers to as *primitive* or *invented*. Naming objects by forming an association between a word and an object would be a characteristic activity of such a language-game. In the second sense, language-games correspond to the naming of objects, the repetition of names, and the activity of word games. Whereas primitive language-games refer to the direct use of language in ways that are presumed to be minimally dependent on custom and convention, language-games of the second sense depend explicitly on a context to convey the designed purpose. Rote learning and the repetition of algorithms and other fully determined activities, such as "invert and multiply" or "the area of a triangle is half the base times the height," illustrate this type of language-game within the language activities of normal mathematics classrooms. The third type of language-game focuses on the whole context of the language activity "consisting of language and the actions into which it is woven" (1991, §7).

Throughout the development of the concept of the language-game, Wittgenstein's intention is to highlight that in our attempt to make sense of words, we always presuppose an already established practice and that taking the correct meaning for these words essentially depends on our ability to appreciate the nature of this activity. Even in examples of ostensive definition in which primitive language-games of pointing and repeating a word are used, the production of meaning does not reside in the association of word and object (or image), but in the knowledge of what to do with this association once it is established. For Wittgenstein, knowledge of this kind belongs to the linguistic domain; it relates to the overall role of the word in language as provided by the grammatical structures of language (1991, §§29–30). Knowledge of this kind is embedded not so much in the structures and logical forms of reality as in grammatical structures of language itself.

CLARITY AS A REQUIREMENT OR AS A GOAL FOR INSTRUCTION?

One important implication flowing from the view of language just expressed relates to the way in which language is used in Wittgenstein's analysis itself. In keeping with his account of the formation of meaning, such key concepts as language-game, grammar, form of life, and so on, are not defined by denotation. Meanings are not obtained directly as though a replacement sets of words could achieve the same end. This feature of Wittgenstein's work clearly challenges the expectations of the reader who is used to working with terms that are supposedly defined in advance of the analytic tasks that they are purported to perform. For Wittgenstein, progress toward understanding meaning can only be indirect. Any deep analysis of language must, he argues, operate at the frontier of what makes sense and what does not, that is, move away from the "purest crystal" (1991, §97) of clear definition toward the "blurred edges" of incipient ambiguity and actual use. Wittgenstein is deeply sceptical about the efficacy of an investigation conducted on the conventional requirement that the terms of an analysis be made clear prior to the investigation itself. Such a requirement, he argues, is like attempting to walk on ice—without friction, no forward movement is possible.

> We have got on to slippery ice where there is no friction and so in a certain sense the conditions are ideal, but also, just because of that, we are unable to walk. We want to walk: So we need *friction*. Back to the rough ground! (Wittgenstein 1991, §107, italics in the original)

"Rough ground" in his view is the domain of actual communicative events taken in their entirety. A principal task of analysis is to investigate the specific and particular, to "lay bare" how language functions in its roughness and diversity. For such a task to be complete, we must "look and see" (1991, §66), rather than speculate. "One cannot guess how a work functions. One has to *look* at its use" (1991, §340, italics in the original). Philosophical investigations must be aimed at making visible the profusion of linguistic differences that both give character to any actual communicative event and, indeed, make it possible. From the perspective of thinking about how communication works, language is too easily smoothed over and, differences removed, idealized. In our first attempts to analyze how meaning is generated, we tend to become trapped by grammatical illusions derived from language-games extrinsic to the language event under analysis.

Focusing on language as action leads Wittgenstein to a further series of far-reaching conclusions of relevance to this study. The first relates to a reappraisal of the nature of the task of rational investigation itself. Whereas in traditional perspectives the task is to solve problems, Wittgenstein's philosophy operates to "dissolve" them (Hunnings 1988). His procedure is to "lay bare" how the problem at hand relies on over-riding features characteristic of the particular context in which the problem arises. For Wittgenstein,

> [p]hilosophy simply puts everything before us, and neither explains or deduces anything. — Since everything lies open to view there is nothing to explain. For what is hidden, for example, is of no interest to us. (1991, §126)

Language, when it is made to operate beyond the domain for which it is tailored (§132), becomes entangled by its own logic, which leads to the "bewitchment of our intelligence" (1991, §109). Against this, the purpose of philosophical investigations must be directed toward developing a recognition and description of the workings of language.

> The fundamental fact here is that we lay down rules, a technique for a game, and that then when we follow the rules, things do not turn out as we had assumed. That we are therefore, as it were, entangled in our own rules. This entanglement in our rules is what we want to understand (i.e., get a clear view of). (Wittgenstein 1991, §125)

In Wittgenstein's terms, to understand this entanglement and how language is implicated in it is equivalent to applying therapy (1991, §133) to the conceptual problems we get into when trying to think our way through problems related to meaning.

A second conclusion relating to the volume of communication focuses on the crucial role played by flexible rather than rigid distinctions among alternative meanings for words in our effort to communicate freely. At first this insight might be taken to state the evident fact that when coming to terms with somebody else's expressions, we often have to adjust our inferences in light of the ongoing interaction. Although Wittgenstein would, of course, not deny this, his point, a much stronger one, is that rational inquiry does not depend on, nor could it exist with, an unambiguously clear articulation of distinct ideas. As a metaphor for rational inquiry, the notion of "clarity," borrowed from visual perception, is never far away. We speak of "clear thinking" and "making oneself 'clear'" and, as we have seen, the idea that pedagogic representations must be transparent is found appealing. Wittgenstein suggests, however, that the use of this metaphor leads to mistakes. For one thing, it erroneously suggests that rational explanations (such as those accompanying efforts to teach mathematics) achieve clarity as a result of their analytic efforts; whereas, he notes, analytic effort, as it is usually described, can operate only by presupposing clarity as a crucial requirement. Clarity is a prerequisite for rational explanation, not a consequence of it. Thus, for Wittgenstein, it is not assumed that reasoning requires accompanying explanations to be clear. Reasoning merely requires that things be cleared up, and being clear is simply one way to achieve clear ends in actual situations.

These conclusions suggest that the ideal of clarity in pedagogic interactions needs to be considered more advisedly. To be sure, it is tempting to think that instruction will proceed smoothly once the seminal concepts associated with particular mathematical ideas are made clear. For instance, in the case of elementary numeration, it might be supposed that addition is most effectively taught once such concepts as number and such processes as grouping are clearly and distinctly comprehended. In essence, however, such a view reflects the rationalist presuppositions that Wittgenstein's thinking finds unconvincing. What then might a Wittgensteinian approach to the problem of clarity within pedagogic interactions be? Although Wittgenstein did not focus on a detailed examination of the pedagogic interaction in his work, it is possible to make the following points, aligned as they are with the general train of his thinking as indicated in *Philosophical Investigations* (1991). For Wittgenstein, instruction moves forward because friction is generated within the pedagogic interaction. Friction, in turn, is possible in situations in which the episode encounters "rough ground." The use of this image is meant to connote the importance of retaining the "blurred edges" to facilitate the ongoing production of meaning within linguistic exchanges. Now this may be interpreted as suggesting that the ideal teacher must purposely confuse a learner by laying bare what is particular within a situation, while leaving the task of smoothing over these differences to the learner. Such an inference would not, of course, be correct, however. Rather, Wittgenstein draws our attention to how language works in ordinary everyday situations. It is in these, replete with their vagueness and diversity, that we find the key to improving our levels of communication with students. As noted above, Wittgenstein recommended paying attention to ordinary language usage as a therapy for the conundrums of modern philosophy; likewise, it is argued, a greater preoccupation of this kind would provide relief from the unproductive tensions that frequently pervade environments for learning. In the following section, these ideas are illustrated with reference to an interaction with students.

ANATOMY OF A PEDAGOGIC INTERACTION

In the following interaction, a number of senior secondary school students are discussing, with the investigator acting in the role of questioner, the work they had previously undertaken with their teacher, identified in the transcript as Ms. Jenkins. The stimulus for the discussion is a video recording of the actual teaching episode, which had taken place the day before. In the part of Ms. Jenkins's lesson related to the discussion transcribed below, students were asked to consider the following question: If z_1 and z_2 are complex numbers, what is arg (z_1/z_2) expressed in terms of arg z_1 and arg z_2? The focus of the transcript is on the use of dummy variables as indices in expressions involving complex algebraic manipulations.

1. *Questioner:* Now the very first step here, where you've got arg (z_1/z_2), Ms. Jenkins wants you to focus on z_1 divided by z_2. Now the first thing that she did was to write that out in a trigonometric form, or a polar form. And she wrote on the top line, what did she write?

2. *Sarah:* r_1, … [inaudible]

3. *Questioner:* Outside of?

4. *Sarah:* Inside the brackets, I think it's $\cos \theta_1 + i \sin \theta_1$.

5. *Questioner:* Why did she say r_1 and θ_1?

6. *Students:* (*Several at once*) Because that's the modulus and argument for z_1.

7. *Questioner:* Sorry, just explain? Sorry, who's talking?

8. *Alan:* Because, well, we've got subscript 1, for z_1, we sort of use the same subscript, probably.

9. *Questioner:* Would it have mattered what subscript? If she'd written '2', would that have been wrong? If she had written r_2 would that have been wrong?

10. *Alice:* Only if she had have, it would have been confusing, because you've got z_1 and z_2, and then you've got, it would be easier to have r_1 and θ_1 then you've got, it makes a link there, so you have, you say that it's with the same, the same problem.

11. *Sarah:* And also, it linked up to what we did yesterday, because we used z_1 and θ_1 and r_1 when we were doing the multiplication as well, so it just tied in with what we did yesterday.

In what follows, an attempt is made to illustrate how language mediated the pedagogic interaction observed. To this end, a framework for analysis has been drawn from Wittgenstein's philosophy of language.

A characteristic of Wittgenstein's approach is that language does not picture mathematical concepts; it helps to constitute them. A focus on language-games directs our attention to the intersection between language as a meaning-making action and language as an aspect of meaning-making that corresponds to the lived purposes, intentions, and contexts of people. One way to illustrate these ideas would be to focus on the question asked (5): "Why did she say r_1 and θ_1?" Students responded to this question in three quite different ways. The first, (6), uses a principle of substitution whereby words and symbols can be replaced by, or substituted for, each other according to rules established by convention. Wittgenstein regarded language-games of this sort as primitive in that they are based on an activity of making direct or ostensive associations between linguistic utterances and their concrete referents. The second response, (8), involves the repetition of names (e.g., the word *subscript* in these lines) and an activity of rehearsing familiar patterns of words and symbols ("we sort of use the same subscript"). The third response, (10), consists of more mathematically specialized utterances. For instance, concepts of difference ("it would have been confusing") and identity ("the same, the same problem") are distinguished, an inferential passage from one to the other is established (note the complex use of if-then statements), and words of linkage (*link*, *linked*) and juxtaposition (*tied in*) are used. Illustrated here, particularly in this third response, is the high degree of interweaving between language and the actions that language facilitates. What these examples begin to show is that mathematical concepts do not exist apart from the language contexts in which they are used. Although all three responses are meaningful within the circumstances in which they are uttered, only the last gives voice to the concept of the dummy variable. However, the concept does not exist apart from this third response; rather, as a form of words it arises as a tailor-made solution to the problem generated by the apparent inadequacy of the students' initial responses. In the face of this inadequacy, students were obliged to explore and develop their range of linguistic maneuvers; each attempt giving rise to new opportunities for interaction. Mathematical concepts were not being pictured by language; instead, language was being used as an activity to get certain things done—in this case, the need to satisfactorily address a particular line of questioning.

As argued previously, thoughts on language-games and sensitivity to the placement of words in ordinary language (grammar) direct Wittgenstein to the conclusion that meanings do not belong to individual words and utterances but to the form of life to which activities belong. It is this point, I think, that Wittgenstein emphasizes in his somewhat cryptic epigram, "If a lion could talk, we could not understand him" (1991, §23). The experience of teachers talking and students not understanding (and vice versa) is, of course, far from being an uncommon cause for concern. If this is accepted, then an important question for successful communication might be stated as follows: To what language-game does a speaker or writer intend an utterance to be associated? In the interaction reported previously, for instance, students were observed to experience various levels of uncertainty relating to their participation within particular language-games. In what follows, then, the role of uncertainty in the progress of the observed interaction is explored.

The first kind of uncertainty to be observed was about how to play a given game. For instance, Alan's explanation (8) was apparently colored by doubt concerning the subject matter he was called on to deal with. Another kind of uncertainty observed concerned the question of how to discern which language-game is pertinent within the terms of a given interaction. As the final "probably" in Alan's response seems to indicate, for instance, his belief was at best tentative concerning the appropriateness of his response on this occasion.

Given these observations, it seems plausible that student responses were not primarily aimed at speaking the truth about the mathematical situation set before them. Instead, utterances were designed to resolve the awkwardness of a social interaction that was perceived, in some sense, to be going wrong. By (6) of the interaction, for instance, the line of questioning adopted had seemed to prove futile; students appeared frustrated by their apparent inability to generate and sustain a pedagogic dialogue acceptable in the context. Moreover, in (7) the questioner seemed unable or unwilling to assist in the task of providing the certainty required to clear up things. Whereas the first utterance of "sorry" in (7) can be thought of as an emphatic refusal to accept the unproblematic legitimacy of the response indicated in (6), the second "sorry," hinting at a transposition of the roles of questioner and student, appeared to call on students to defend their response and take a stand in relation to their own position within the ensuing dialogue. Although asking students to declare their hand, the questioner seemed disinclined to reciprocate and thereby appeared to undercut the conventional belief that a teacher's role is to vouchsafe students' passage to mathematical competence (cf. Brousseau 1986, p. 288).

Crucial to the episode discussed in the previous paragraph was the questioner's success in challenging students' adherence to certain ways of expressing the terms of a particular mathematical situation. New activities with words were seen to arise as a consequence, and these were later associated with the conceptual development of mathematical notions. However, meaning, unclear and uncertain though it may have been for most of this interaction, was not obtained by merely replacing one set of words with another. Nor was meaning obtained by marking off the boundaries of a concept, as if it were detachable from the words and context in which it arose. Instead, meaning was firmly fixed to a particular interaction within which it became realized as an activity. In elaborating on this notion, Wittgenstein comments in the following way:

> Frege compares a concept to an area and says that an area with vague boundaries cannot be called an area at all. This presumably means that we cannot do anything with it. But is it senseless to say: 'Stand roughly there'? Suppose that I were standing with someone in a city square and said that. As I say it I do not draw any kind of boundary, but perhaps point with my hand—as if I were indicating a particular spot. (Wittgenstein 1991, §71)

Thus for Wittgenstein, in an attempt to make meaning of an interaction, the task will characteristically involve looking and seeing to ascertain how utterances are placed within the overall manifold of language and action; decoding language by applying references to otherwise remote objects will not suffice. If approximation, even vagueness, is to be seen as an imperfection within the linguistic domain, then it must be concluded that key aspects of successful communication lie beyond the purest crystal of exposition. As a means of mediating communication, language is not neutral. At worst it distorts the formation of ideas; at best it makes the construction of concepts and communication possible.

FINAL REMARKS

Analyses in this chapter have focused on language to illustrate the argument that communication is achieved when parties collaborate in an activity of joint concern. Meaning cannot be abstracted from the particular circumstances in which it is formed. Language, it has been argued, has a double-sided character. Whereas in some circumstances it assumes the character of a representational code, its operations are far from exhausted by this role. In that it gets things done, language revolves around technical concerns; yet it also constitutes the objects about which it speaks, and for this reason, language transcends all such overly reductive characterizations. These conclusions were found to have important implications for the role that clarity plays in mathematics teaching. Assumptions regarding the value of clear instruction were shown to be less self-evidently valid than is often supposed. The study opens an avenue for the detailed examination of the language of pedagogy in mathematics, which promises to deliver insights of practical advantage to teachers.

What, then, can be said about the issues for mathematics education that arise from the view of language advanced in this work? In responding to this question, we need to distinguish two kinds of pedagogic analyses. The first type assumes that language is nonrepresentational and would seek to recover the diversity of language practices relating to meaning-making projects, such as those engaged in by teachers and learners. Such analyses are aligned with contemporary postmodernist theory (Lyotard 1984; Lyotard and Thébaud 1985; Harvey 1989) and highlight concepts of fragmentation and disjunction within classroom interactions, tending to treat these as ends in themselves. The second type of pedagogic analysis builds on the first, but supplements it with a normative logic. The work of Habermas, in particular his theory of communicative action (1984), provides a rich example of this second kind of analysis. Those who favor the first type will be impressed by the Wittgensteinian approach presented in this chapter (see Benhabib [1990]). Such an option would be consistent with a conservative view of social values. Those who favor the second will seek to extend Wittgenstein's analyses to explore systematically the relationship between the production of meaning, value, authority, and privilege within language-games and the forces that organize them on the social plane. This approach would be consistent with a critical or socially transformative agenda for change. In my opinion, a central issue of concern relates to the choice between these types of analyses. But this choice can be made only from the standpoint of what field of meaning or form of life we wish to associate with the concept of the pedagogic interaction.

REFERENCES

Austin, J. L. *How to Do Things with Words.* Oxford: Oxford University Press, 1980.

Bakhtin, M. "The Problem of Speech Genres." In *Speech Genres and Other Late Essays,* edited by C. Emerson and M. Holquist, pp. 6–102. Austin, Tex.: University of Texas Press, 1986.

Barthes, R. *Elements of Semiology.* New York: Hill & Wang, 1968.

Bauersfeld, H. "'Language Games' in the Mathematics Classroom: Their Function and Their Effects." In *The Emergence of Mathematical Meaning: Interaction in Classroom Cultures,* edited by H. Bauersfeld and P. Cobb, pp. 271–92. Hillsdale, N.J.: Lawrence Erlbaum Associates, 1995.

Benhabib, S. "Epistemologies of Postmodernism: A Rejoinder to Jean-Francois Lyotard." In *Feminism/Postmodernism,* edited by L. J. Nicholson. New York: Routledge, 1990.

Bloor, D. *Knowledge and Social Imagery.* Boston: Routledge & Kegan Paul, 1976.

———. *Wittgenstein: A Social Theory of Knowledge.* New York: Columbia University Press, 1983.

Brousseau, G. "Basic Theory and Methods in the Didactics of Mathematics." In *Second Conference on Systematic Cooperation between Theory and Practice in Mathematics Education,* edited by V. F. L. Verstappen, pp. 109–61. Lochem, Netherlands: 1986.

Cobb, P. "The Tension between Theories of Learning and Instruction in Mathematics Education." *Educational Psychologist* 23 (1988): 87–103.

———. "A Constructivist Alternative to the Representational View of Mind in Mathematics Education." *Journal for Research in Mathematics Education* 23 (1992): 2–33.

Confrey, J. "Concepts, Processes, and Mathematics Instruction." *For the Learning of Mathematics* 2 (1981): 8–12.

Derrida, J. *Of Grammatology.* Baltimore, Md.: Johns Hopkins University Press, 1976.

Dienes, Z. *An Experimental Study of Mathematics Learning.* London: Hutchinson, 1963.

Ellerton, N., and M. Clements. *Mathematics in Language: A Review of Language Factors in Mathematics Learning.* Geelong, Victoria, Australia: Deakin University Press, 1991.

Ellis, J. *Against Deconstruction.* Princeton, N.J.: Princeton University Press, 1989.

Ernest, P. *The Philosophy of Mathematics Education.* London: Falmer Press, 1991.

Foucault, M. *The Archaeology of Knowledge.* London: Tavistock, 1972.

Habermas, J. *The Theory of Communicative Action.* Boston: Beacon Press, 1984.

Hamlyn, D. W. "Education and Wittgenstein's Philosophy." *Journal of Philosophy of Education* 23 (1989): 213–22.

Harvey, D. *The Condition of Postmodernity.* Cambridge, Mass.: Blackwell, 1989.

Holquist, M. *Dialogism: Bakhtin and His World.* London: Routlege, 1990.

Hunnings, G. *The World and Language in Wittgenstein' s Philosophy.* Albany, N.Y.: State University of New York Press, 1988.

Kanes, C. "Language, Speech, and Semiosis: Approaches to Postconstructivist Theories of Learning in Mathematics." In *Contexts in Mathematics Education: Proceedings of the Sixteenth Annual Conference of the Mathematics Research Group of Australia,* edited by B. Atweh, C. Kanes, M. Carss, and G. Booker, pp. 369–74. Brisbane, Australia: Mathematics Education Research Group of Australasia, 1993.

———. "Using Mathematics Ideas and Techniques." In *Learning in the Workplace: Tourism and Hospitality,* edited by J. Stevenson, pp. 51–111. Brisbane, Australia: Centre for Skill Formation Research and Development, 1996a.

———. "Investigating the Use of Language and Mathematics in a Workplace Context." In *Technology in Mathematics Education: Proceedings of the Nineteenth Annual Conference of the Mathematics Research Group of Australia,* edited by P. Clarkson, pp. 314–21. Geelong, Victoria, Australia: Deakin University Press, 1996b.

———. "An Investigation of Artifact Mediation and Task Organisation Involving Numerical Workplace Knowledge." In *Good Thinking: Good Practice: Proceedings of the Fifth International Conference on Post-Compulsory Education.* Brisbane, Australia: Centre for Learning and Work Research, 1997a.

———. "Towards an Understanding of Numerical Workplace Numeracy." In *People, People, People: Proceedings of the Twentieth Annual Conference of the Mathematics Education Research Group of Australasia,* edited by F. Biddulph and K. Carr. University of Waikato, Mathematics Education Research Group of Australasia, 1997b.

———. "Exploring the Nature of Numerical Operations in Workplace Settings." In *Airline Ticketing Operations: Knowledge at Work,* edited by F. Beven. Brisbane, Australia: Centre for Learning and Work Research, in press.

Kristeva, J. *The Kristeva Reader.* Oxford: Basil Blackwell, 1986.

———. *Language the Unknown: An Introduction to Linguistics.* New York: Columbia University Press, 1989.

Lacan, J. *Ecrits: A Selection.* New York: Norton, 1977.

Lyotard, J.-F. *The Postmodern Condition: A Report on Knowledge.* Minneapolis: University of Minnesota Press, 1984.

Lyotard, J.-F., and J.-L. Thébaud. *Just Gaming*. Minneapolis: University of Minnesota Press, 1985.

Norris, C. *The Deconstructive Turn: Essays in the Rhetoric of Philosophy*. London: Methuen, 1983.

Pimm, D. *Speaking Mathematically: Communication in Mathematics Classrooms*. London: Routledge & Kegan Paul, 1987.

Resnick, L. "Towards a Cognitive Theory of Instruction." In *Learning and Motivation in the Classroom*, edited by S. G. Paris, G. M. Olson, and H. Stevenson. Hillsdale, N.J.: Lawrence Erlbaum Associates, 1983.

Ryle, G. *The Concept of Mind*. London: Hutchinson, 1971.

Saussure, F. de. *Course in General Linguistics*. London: Duckworth, 1983.

Staten, H. *Wittgenstein and Derrida*. Lincoln, Neb.: University of Nebraska Press, 1986.

Strawson, P. *Logico-linguistic Papers*. London: Methuen, 1971.

von Glasersfeld, E., ed. *Radical Constructivism in Mathematics Education*. Dordrecht, Netherlands: Kluwer Academic Press, 1991.

Vygotsky, L. S. *Mind in Society: The Development of Higher Psychological Processes*. Cambridge: Harvard University Press, 1978.

———. *Thinking and Speech*. New York: Plenum, 1987.

Walkerdine, V. *The Mastery of Reason: Cognitive Development and the Production of Rationality*. London: Routledge, 1988

———. "Reasoning in a Post-Modern Age." In *Mathematics, Education and Philosophy: An Instructional Perspective*, edited by P. Ernest. London: Falmer Press, 1994.

Watson, H. "Language and Mathematics Education for Aboriginal-Australian Children." *Language and Education* 2 (1988): 256–73.

———. "A Wittgensteinian View of Mathematics: Implications for Teachers of Mathematics." In *School Mathematics: The Challenge to Change*, edited by N. Ellerton and M. A. Clements, pp. 18–30. Geelong, Victoria, Australia: Deakin University Press, 1989.

Wertsch, J. *Voices of the Mind: A Sociocultural Approach to Mediated Action*. Hemel Hempstead, England: Harvester Wheatsheaf, 1991.

Wittgenstein, L. *Tractatus Logico-Philosophicus*. Paris: Gallimard, 1961.

———. *Remarks on the Foundations of Mathematics*. Oxford: Basil Blackwell, 1967.

———. *Philosophical Investigations*. Oxford: Basil Blackwell, 1991.

Part 3

Different Styles and Patterns of Communication in the Mathematics Classroom

7

Pupil Language–Teacher Language: Two Case Studies and the Consequences for Teacher Training

Albrecht Abele

Pädagogische Hochschule, Heidelberg

DIFFICULTIES in understanding are frequently held responsible for the failure of pupils of all ages in mathematics lessons. The pupils' standpoint—and very often that of the parents—can be characterized by such formulations as "The teacher can't explain properly." The teacher's standpoint, however, can be characterized by statements following this pattern: "Pupil X simply cannot understand the subject matter regardless of how often I explain it." Although each standpoint blames something different, both proceed from the concept of a two-phase basic structure in the mathematics lesson, in which language—as "explaining" on the part of the teacher and as "understanding" on the part of the pupil—is regarded as the foundation of the mathematical learning process. In contrast, we have the idea of learning by discovery and recreating inventive experience, promoted by Freudenthal (1973, p. 116) and others, which takes into account the process of forming concepts through doing, as called for in primary education in accordance with educational psychology. Thus there is a need to redefine the role of language and its significance for mathematics lessons in primary school.

Language, including colloquial language used by pupils and teachers, has a two-fold function in mathematics lessons:

1. Mathematics itself has a linguistic form. The symbolic character of the so-called mathematical language, as the most striking characteristic, often conceals its grammatical and semantic structure. Pupils and teachers frequently understand mathematical formulas as something nonlinguistic, which leads to methodological problems (Abele 1988).
2. Language is the general instrument of communication and, therefore, the medium for talking about mathematics in class. The analysis of communication processes in mathematics lessons is consequently an important, urgent, and long-neglected challenge for research in the field of methodology in mathematics teaching.

This chapter describes two experiments concerning the same mathematical problem (a comparison of speed) in a third- and a fourth-grade lesson. The lessons were taught in two third- and fourth-grade classes by different teachers, immediately before the students were given an examination. Both teachers used language as part of a specific methodological design: the first, a teacher-led discussion and the second, a small-group session with no input from the teacher. By comparing the two lessons we observe—

- the importance of the structure of classroom interaction;
- the advantages of the second situation, in which the authority of the teacher is superseded by the authority of the subject;
- some of the conditions that are the necessary elements of the learning-teaching process and that depend on the language in the mathematics classroom.

"YOU DON'T UNDERSTAND ME PROPERLY"

This is the cry of resignation of a student teacher who feels deceived in her expectations as a teacher. In what context do we observe this utterance (Abele 1984; Videodocumentations 1984, Transcript 84061k)?

111. *Teacher:* Yes, but, the question was whether you can work it out like this, too, if you now take, for example, a certain period of time for both. For example, how many kilometers both travel in 5 minutes or in 10 minutes.
112. *Pupil:* In 10 minutes.
113. *Teacher:* Can it be calculated in that way as well?

114. *Pupil:*   You can't do that exactly.

115. *Pupil M:* In 10 minutes, then it would be 30 divided by 3, that ... then it would be 30 minutes (for 10 kilometres) because ... [inaudible]

116. *Pupils:*   (2 hours) divided by 4 are also 30 minutes.

117. *Teacher:* Now again you haven't worked out the minutes, but in per 10 kilometers! You don't understand me properly! You've given—so that you can compare it—you've given the same number of kilometers for both.

118. *Pupil:*   Hm.

119. *Teacher:* Either 1 kilometer, or as you just said: 10 kilometers. And now we've got 2 different time designations as well, and if you could now work out how many kilometers Peter cycles in, for example, 10 minutes and then still work out how many kilometers Jens cycles in 10 minutes. Would that work, too?

120. *Pupil:*   But that's already ..., that's the same, they are equally fast.

121. *Teacher:* No, but can you also work it out that way round? How long each cycles in half an hour?

122. *Pupil 1:*   Yes.

123. *Pupil 2:*   Well, actually you could.

124. *Pupil 3:*   Actually, yes.

This passage from Abele (1988) indicates, not only superficially, a certain lack of comprehension or, more precisely, a mutual lack of understanding, but proves yet again how very common and complete this above-mentioned misconception of mathematics lessons is among teachers (and consequently also among pupils). This conception is seen more frequently not in such direct statements as, "You don't understand me," but rather in the teacher's question, "Have you understood that?" It is seen more subtly in the teacher's idea, which is usually not expressed, that the pupils have understood him or her after this thorough explanation and that they can go on to do further exercises.

A conversation takes place when two or more people feel the need to discuss events, to exchange opinions, or to tell others about their own experiences and knowledge gained. This is also true in mathematics lessons. Using language is not an end in itself, but a means to an end, communicating thoughts, facts, sequences of events, ideas on solving problems, and so on. Typical situations giving rise to conversations that enable the pupil to report on his or her own mathematical experience or that expect individual pupils or groups of pupils to make decisions on their own occur too rarely in the normal mathematics lesson. Methodologically, an expansion of the structurally narrow conversation network of teacher's question and pupil's answer is called for, at least through offering questions with several ways of reaching the correct solution to create new, extended possibilities for conversational situations.

This happened in the example given, and our criticism begins at the point where the teacher tries to guide the pupils along a second path to solving the problem. But first let us consider the problem on which the children in this scene were working:

> Four boys are riding their bikes:
>
> Uwe cycles 30 kilometers in 2 hours.
>
> Peter cycles 30 kilometers in 1 hour 30 minutes.
>
> Jens cycles 40 kilometers in 2 hours.
>
> Bernd cycles 40 kilometers in 1 hour 30 minutes.
>
> Which of them is the fastest?

First of all, we realize that two basically different approaches to the problem are possible:

(1) The one who covers the same distance in the shorter time is fastest (race model).

(2) The one who covers the greater distance in the same time is fastest (the number of kilometers per hour [km/h], speedometer model).

The numbers have been chosen so that the first strategy leads directly to the solution in two cases (Uwe/Peter and Jens/Bernd), whereas the second strategy for solving the problem gives an immediate answer in two other cases (Uwe/Jens and Peter/Bernd). Only in the decision between Peter and Jens do we have to call on the services of arithmetic. Appropriate guiding questions might include these:

- How long does it take each boy to cycle 10 kilometers?
- How far does each travel in half an hour?

Contrary to the expectations of the teacher, however, the pupils are not prepared to consider both possibilities to the same degree. They quite clearly prefer the point of reference suggested by the first method of finding the solution. The central quantity for the pupils is always the distance. Proof is to be found in these comments from the pupils:

"Because it takes him [the fastest] the least time to cover the longest distance."

"Because it takes him [the slowest] 2 hours ... um ... for ... for a slow [sic] distance, that is longer than the one who cycles the faster [sic] one."

"So the one who takes longer is slower" [if they cycle the same distance].

"And that one ... and that one's got ... the other one's got a longer distance and shorter time. So that means he's ... faster."

Their personal experience with races forms the framework for the pupils' way of thinking. They can tell, even without a stopwatch, who is the faster of two runners. It is the one who crosses the finish line first. It takes him the shorter time to cover the distance.

The teacher, however, obviously takes her central point of reference from the speed concept, "distance per unit of time," with the dimension km/h. That is, she gives priority to the conceptual framework "speedometer" for solving the problem. As the conversation in class continues, it is emphasized that Uwe must be the slowest and Bernd, the fastest. The pupils find that their race calculations provide sufficient proof for the conclusion that Jens and Peter cycled at the same speed. The teacher is obviously willing to accept this solution as only preliminary.

The passage quoted from the conversation shows clearly how much the teacher's concept in her lesson planning differs from the actual way in which pupils approach solving problems. The solution she had planned, as can be reconstructed from the record of what actually happened in the lesson, would have to be accomplished in two steps. In the first step, the pupil would have to come to the conclusion that a direct comparison in this case is not possible. In the second step, a common unit of reference—"5 minutes" or "10 minutes" or even "half an hour"—would first of all have to be established to determine the result completely.

The pupils react defensively in two respects when faced with the plans the teacher has in mind. The teacher's statement that "if you just look at it, they can't be compared directly" is dismissed as irrelevant by pupil A. He interrupts the teacher vehemently with the objection "Yes, but if you do a few sums then, um, you need a few sums to find out...."

He obviously finds the teacher's systematic orientation unimportant. The statement as such, that the two cannot be directly compared, is of no consequence as far as he is concerned because indeed he can obtain the desired result with "a few sums," as happened just before. During this phase of the discussion, the teacher clearly remains bound by her systematically oriented, methodical thinking, whereas the pupils put all their effort into arguing about content to answer the given question, for example, "They are the same because, in one respect, Jens has won because he cycles farther ... (agreement from other pupils) ... and, in another respect, Peter has won because he takes a shorter time." Thus the teacher also uses a framework different from that used by the pupils as regards the strategy for finding the solution. As far as the teacher is concerned, we could talk of a framework that is systematically oriented and has become relatively rigid because of the planning. As far as the pupil is concerned, we could instead talk of a framework that is characterized by the idea: "In mathematics lessons I get problems that can be solved by doing sums."

The idea of a solution, still seen in qualitative terms—longer distance–shorter time—is taken up by another pupil, and in the final part ("Peter won because he's got a shorter time") it is carried on in quantitative terms: "So you can do it divided by 30, then, you see, it would be 3 minutes ... per kilometer." The teacher is very hesitant in accepting the content of this partial solution and tries repeatedly to force the pupils into her own analytical, systematic position by means of rather confusing questioning. The pupil, though, characterizes the result on the level of his own calculations. When the teacher asks, "Right then, what have you done?" he answers in the only way possible for him with a statement that, however, does not answer the question the teacher asked: "It took them both 3 minutes [for 1 km]." Apparently the teacher does not deem this adherence to content in the pupil's answer to be of importance, since she merely comments, "So, if you just look at it they can't be compared directly. You need ... (interruption by the pupils) ... a quantity common to both." With this comment the teacher introduces another clue, "[...] what about taking, for example, the same time for both?" which leads to the method of solving the problem wanted by the teacher.

The pupils, however, continue to prefer their old method within the race framework, which is refined further and also varied. They now compare the times it takes Jens and Peter, respectively to cycle 1 kilometer and, afterward, also 10 kilometers. Another three attempts on the part of the teacher, which are progressively reduced to "yes/but" instructions, are necessary to make the pupils relinquish their own ideas about how to solve the problem and be prepared to accept the (teacher's) framework characterized by the speed concept. See lines (111), (117), (119), and (121).

When the result "10 kilometers in half an hour" has been established, the teacher asks, "How did you work that out?" Several pupils join in the answer: "I, I divided.... I just divided by 3." Others voiced agreement with, "Right!" After the teacher's supplementary question, "Why did you divide it by 3?" another pupil answers, "Because it takes, um, 3 minutes for 1 kilometer." It could not have been demonstrated more clearly that in spite of the teacher's efforts, the pupils—at least partially—could not be induced to depart from their original conceptual framework. They were prepared to think along the lines of the other method again only after further specific, guiding questions, such as, "How many half hours are there in 1 hour 30 minutes then?" The race framework, with distance as the common quantity of reference, proved to be firmly set in the minds of the pupils even as the lesson continued.

The differences in emphasis in the process of solving the problem can also be noted as a striking result of the analysis.

- The pupil strives toward the solution itself as directly as possible.
- The teacher attaches more importance to the structuring of the process involved, which is beyond the comprehension of the pupil at this point.

THE PUPIL JURY

The second scene is taken from videotape 85081a (Abel 1986b; Videodocumentations 1986) about the same problem. Only the names of two cyclists were changed:

Uwe	30 km	2 hrs
Heike	30 km	1 hr 30 mins
Fred	40 km	2 hrs
Bernd	40 km	1 hr 30 mins

When the scene is analyzed, the following characteristics are particularly noticeable. The student teacher sets up the group of pupils as a jury, which has to make an unequivocal decision about the order of the four cyclists on the basis of speed. In this way the teacher declines to take on the frequently (and readily) accepted role of judge between right and wrong solutions offered by the pupils. The four pupils are left with the task of making this decision and consequently have to give reasons for and defend their individual solutions. These reasons become negotiating positions, with arguments, until a consensus—the solution—has been achieved. As a natural consequence of this conceptual approach, the teacher leaves the room after she has presented the problem to the pupils, allowing them to solve it on their own. This was the first phase of the lesson.

In the second phase of the lesson, the result, which was at first accepted by the majority, but which is really wrong, is discussed again —this time in the presence of the teacher—until finally the correct solution gains acceptance by force of argument in the third phase.

This conversation among the pupils, lasting about five minutes, deserves an in-depth analysis of the pupils' possibilities of argumentation about content, also, but this would go beyond the scope of this chapter. Here the role of the teacher as a silent observer is emphasized. She makes only five, out of eighty-three, contributions to the discussion, controlling the formal aspect. These contributions by the teacher follow:

"Yes?"

"Who then?"

"You are to form a jury to establish who is actually first, second, third, and fourth."

"You can't quarrel; as judges you have to come to one decision."

"Yes, you must convince him."

In the first phase (the pupils establish the order according to speed), the two middle positions are wrongly placed as numbers 2 and 3 (instead of both as 2), although the pupil P1 had already found the correct answer at a very early stage and even told her fellow pupils. This mistake made by the group is at first surprising for the observer, and the question arises why pupil P1 had no success with her opinion.

As a rule it can be assumed that information is accepted in a group either if it is immediately plausible to everyone or if it is given by a socially dominant member of the group. Obviously neither is the case here; the video documentation makes this quite clear.

Several explanations are possible for why pupil P1's solution, in its factual context, was not immediately plausible to her three fellow pupils. The pupils had the unexpressed (intuitive) hypothesis, or at least expectation, that all four competitors cycle at different speeds. Even pupil P1, with her correct solution, follows her reason, "It would take Heike 30 minutes for 10 kilometers and Fred too [...] 30 minutes [for 10 kilometers]," with the additional comment, "Then that can't be right, actually." This formulation is used again and again by pupils, in turn.

Other approaches to a solution—also wrong—compete with the correct solution. Finally a suggestion is accepted that differs least from the general expectation of the pupils. At the end, the pupil who gave the reason for this solution is the most difficult to convince that the other solution is correct.

The explanation of the solution that P1 is able to formulate is not quite complete and is linguistically awkward so that her fellow pupils do not understand what she means:

76. *P1:* Wait a minute ... 30 kilometers ... 30 kilometers ... in 2 hours 40 kilometers, in 2 hours. Then it would have taken 30 minutes for 10 kilometers. So 30 minutes (for 10 kilometers), in 30 minutes 10 kilometers, for Fred ... 30 kilometers (in 1 1/2 hours) and Heike also needed 30 minutes (or 10 kilometers).

77. *P4:* No, 1 hour 30 minutes.

78. *P4:* I think this one (*indicates Fred*) is actually ... no, this one (*indicates Heike*), this one here is third.

The dispute among the pupils about the second place ends when the outwardly concurring opinion is presented: "So Heike is actually second." Pupil P1 has also given up contradicting for the time being and is obviously waiting for the teacher to appear to get her opinion accepted after all with the backing of the teacher's authority. When the teacher enters the room, P1 at once unexpectedly opens the discussion with her opinion, which differs from the result she has just tolerated: "Really they both would have had second place, you know." She counters the contradictions of her fellow pupils, "No," "No, Heike was faster," with arguments, "But I think, that with 1 hour 30 minutes [time taken by Heike for 30 kilometers] ..., it would also take [...] 30 minutes for the 10 kilometers. It would actually take Fred here 30 minutes for the 10 kilometers, too; because 2 hours here but then it's 10 kilometers more.... It took half an hour longer. Both are really second." The opposite standpoint of P2, expressed with an inquiring glance at the teacher, "No, Heike ... actually," also seeks confirmation from the teacher. However, the teacher gives the role of judge, which the pupils normally considered hers, back to the pupils with the question, "Who then?" And, in this way, the discussion, characterized by factual arguments, arises once more among the pupils with the intention of ensuring that the result of the discussion is the correct solution. The decision of the teacher not to misuse her authority in questions of content, which the pupils can answer independently, inevitably leads the pupils back to factual arguments. Prejudices and previously expressed opinions are reviewed, ideas and conceptions are compared with given data, and different intellectual approaches and methods for finding a solution are brought into play by way of substantiation, until finally the last of the four pupils is also convinced that the solution that is then presented is correct: "I think there are two second places and no fourth." This correct solution is now no longer accepted as one put across by a socially dominant person but is reached by the pupils themselves in a contest of opinions through argument and counterargument on the basis of the given data.

FINAL REMARKS

What conclusions can be drawn from the observation of such situations? The first case study, which could also be described as a "framework conflict" (Krummheuer 1982), and the second study, which can be interpreted as a hidden cognitive conflict in the resolution of which, at first, social components played a too great, and perhaps unexpected, role, draw our attention to the significance of some nonmathematical factors in mathematics teaching. It follows that we should call for structure in the desired communication between pupils and teacher as well as among the pupils. It is necessary that the social interaction in the course of the lesson be seen more clearly in the planning of lessons—even mathematics lessons.

One of the first logical steps to be taken in teacher training is to make new teachers conscious of their own behavior so that they can develop methods of taking this behavior into account in their lesson planning and even changing it. The part of the discussion we saw in the first scene was, after all, bound to fail because the pupils had not yet acquired the routines of classroom dialogue in lessons, observed by Voigt (1984, pp. 161–86, particularly pp. 175–76), which unfold step by step through the teacher's questions—routines on which the teacher unconsciously reckoned. Pupils resist the reduction of language necessary in such a form of dialogue, in which it is not coherent ideas about solutions that are required but only associated terms approximating teacher expectations and combined in more or less stereotyped language patterns.

In no way, however, can it be a question of building up such routines as quickly as possible, which in this context are to be judged negatively.

It must rather be the aim of teacher and in-service training to ensure that teachers build a comprehensive repertoire of methods for leading a discussion and then make use of these methods consciously and not only as a matter of routine. In our example, the attention of the pupils could indeed have been easily drawn to the second way of solving the problem by means of a slight modification in the way in which the problem was presented.

Discussion is triggered not by constantly talking about the subject matter—as was attempted in scene 1—but more appropriately by suitably presenting the problem. This (especially in the first two primary classes) will frequently trigger a reaction among the pupils in the form of action as opposed to verbal efforts.

The second scene provides an example of how the dominance of social constellations in the group can be dismantled in favor of an "authority of the subject matter" through the skillful way in which the teacher reacts.

We can learn from these two examples and from the continuations of the lessons, which are not mentioned in this chapter, that independent discussion by the pupils, without hints by a teacher, leads to a better understanding of the concept of speed than does the teacher-led discussion.

Which consequences and possibilities—and also which difficulties—ensue for everyday mathematics teaching have to be discussed separately. An initial approach has been attempted in Abele (1988). Here it must suffice to draw attention to the extensive methodological repertoire that research in the field of teaching methods in mathematics puts at the disposal of interested teachers.

REFERENCES

Abele, A. "Transkripte 1984." Pädagogische Hochschule Heidelberg, 1984.

———. "Problemlösen in der Kleingruppe." In *Kooperatives Lehren und Lernen in der Schule*, edited by A. Weber, pp. 345–64. Heinsberg: Dieck, 1986a.

———. "Transkripte 1985/86." Pädagogische Hochschule Heidelberg, 1986b.

———. "Kommunikationsprozesse im Mathematikunterricht." *Mathematische Unterrichtspraxis* 5 (1988): 23–30.

Freudenthal, H. *Mathematics as an Educational Task*. Dordrecht, Netherlands: D. Reidel, 1973.

Krummheuer, G. "Rahmenanalysen zum Unterricht einer achten Klasse über 'Termumformungen.'" In *Analysen zum Unterrichtshandeln* (Untersuchungen zum Mathematikunterricht Bd. 5, pp. 41–103, Institut für Didaktik der Mathematik, Bielefeld). Cologne: Aulis, 1982.

Videodocumentations Nr. 84061k Geschwindigkeit (4. Schuljahr), Nr. 85081a Geschwindigkeit (3. Schuljahr). Pädagogische Hochschule Heidelberg, Audiovisuelles Zentrum, 1984/85.

Voigt, J. "Der kurztaktige, fragendentwickelnde Mathematikunterricht—Szenen und Analysen." *Mathematica Didactica* 7 (1984): 161–86.

8

Teacher-Student Communication in Traditional and Constructivist Approaches to Teaching

Maria Luiza Cestari

Kristiansand, Norway

N THIS chapter, the teacher-student communication in the classroom is studied from a sociocognitive perspective and in a sociocultural context. The theoretical background concerning the construction of knowledge, outlined by Vygotsky (1978, 1988), Rommetveit (1974, 1992), Hundeide (1988, 1992), Perret-Clermont (1978, 1991), Perret-Clermont, Perret, and Bell (1992), and others, offers alternatives to Piagetian studies. The aim is to contrast patterns of communication in traditional and constructivist models of teaching.

The inductive-descriptive nature of this research requires special methodologies. To collect data, mathematics lessons were observed in the classroom, audiotaped, and transcribed into protocols. The analysis of these data is inspired by conversational analysis (Taylor and Cameron 1987) and discourse analysis (Coulthard 1990) and to an even greater extent by the dialogical approach to communication (Marková and Fopp, 1990; Wold 1992; Linell and Marková 1993).

Two of the problems facing Brazilian education today are the high drop-out rate and the failure of children to be promoted to the next grade level (Carraher, Carraher, and Schliemann 1985, 1988; Patto 1991). According to the Brazilian Constitution, it is compulsory for all

This research is part of a larger study, initially supported by the National Council of Research (CNPq).

children between seven and fourteen years of age to attend school. However, for every hundred children registered for their first year of school, only fifty-one are still registered in their third school year (Bishop and Pompeu 1991, p. 136).

Although we could blame this failure on the intelligence, stimulus, and motivation of pupils and their families or on the organization of society (Lautrey 1984; Patto 1991), we decided to observe more closely what happens in the classroom, the place where education is actualized. More specifically, we took a closer look at how knowledge is being elaborated through the communication between the teacher and the students.

Traditionally, ideas underlying mathematics teaching in Brazil are quite authoritarian and are based on behaviorist principles. In Recife, Brazil, the project "Aprender Pensando" (Carraher 1989) introduced in 1981 a more constructivist approach to mathematics learning. Part of this project was a constructivist teacher-education program. One aim of the wider frame of study is to compare the classroom communication of three teachers representing both approaches to mathematics teaching as practiced in Recife.

These approaches to teaching are characterized by the following four aspects:

- *Form of communication:* In the traditional approach, the teacher makes more frequent use of statements and orders that are based on his or her own aims, whereas the teacher in the constructivist approach lets him- or herself be guided by the questions that arise in the course of the lesson and the answers given by the pupils.

- *Cognitive strategies:* In the traditional approach, the tendency among students is to memorize the content by imitating a model proposed by the teacher and to perform a repetition of actions until the right answer is obtained. In the second approach, pupils are more involved in reflection and comprehension, thinking about the problems and searching for the answers in their internal repertoire of knowledge.

- *Execution of the task:* In the traditional instruction, teachers tend to emphasize the final product of the task, whereas in the constructivist approach, the importance lies also in the process of the product's realization.

- *Status of mistakes:* This status is totally different in the two approaches. In the first approach, mistakes are synonymous with failure or defeat; teachers are aiming for learning without mistakes. In the constructivist approach, mistakes are considered "perturbations of the milieu," provoking "sociocognitive conflicts" (Perret-Clermont 1978) and "cognitive disequilibrium," which is a suitable condition for learning (Inhelder, Sinclair, and Bovet 1974).

These characteristics are well illustrated throughout the entire study of the protocols. To some extent, they are illustrated in the episodes presented here. To provide some insight into the forms and mechanisms of this particular type of communication between the teacher and the pupils, we have used observational methods, inspired by the ethnographic tradition in research (Garfinkel 1967).

THE SUBJECTS

Two kindergarten classrooms are involved. The teacher in one class has received pedagogical training based on the traditional approach to learning, which emphasizes memorization; the teacher in the other has participated in a project called "To Learn through Thinking," which is based on the constructivist approach that emphasizes comprehension instead of memorization. The principal aims of this project are to suggest to the teacher "to teach through thinking and not to repeat mechanically the steps of a determined teaching methodology; to understand the point of view of the child, to identify which questions can bring her to new discoveries, to propose these questions and wait until she discovers the solution—this is the essence of teaching and learning through thinking" (Carraher 1989, p. 9; translation by the author).

RESULTS

The following are some empirical examples with episodes taken from the protocols. One *episode* is understood as all the exchanges characterized by a thematic continuity, usually with no changes of participants or social roles (Garvey 1984). These episodes are part of a "speech event," a particular type of encounter in a specific place and time, in which objectives are conventionally organized, such as a classroom lesson.

Example 1: The traditional approach

Task: To copy the number 3 on a paper with lines indicating where to write it (the number 3 is already written at the beginning of the line as a model)

Procedure: Children are seated at their tables. The teacher explains the task; afterward she observes and makes comments about their work. *Note:* In Brazil, children refer to their teacher as "Aunt."

Episode 093

1. *Pupil:*   Look at mine, Aunt!
2. *Teacher:*   It's not right, look, it's not right; trace it over and over, and then you will know how to do the 3 below.

Episode 098

1. *Pupil:*   I've already done that, Aunt...
2. *Teacher:*   OK, if you've already traced it and cannot get it, trace it again until you get it right.

Episode 134

1. *Pupil:*   Aunt, I've finished, but I can't do it....
2. *Teacher:*   Oh! you can't do it? Well, go back, try again, trace the dots again, trace over and over, once, twice, three times, as many times as you need to be able to draw it below.

In these three episodes, to deal with the pupils' difficulties in writing the target number correctly, the teacher suggests just one strategy, namely, repeating the action. In episode 134, after having advised the pupil to write the number many times, she indicates that her goal is to have the pupil perform the task correctly.

Episode 174

1. *Teacher:*   Look, place your pencil on the little dot, then, from the little dot you will start to draw the little curves of the 3. If you don't get it right, trace it again, trace it over once, twice, three ti..., as many times as you need to, trace it over.... No Eri, it's not like this, Eri, it's not, look, see, see, let's go here, one turn, another turn, isn't it? You can see it looks like a belly (*barrigudinho*) can't you? Then it must have one belly, and another belly. You must do like this, start from the little dot here, look, see if you trace over, you will get one belly, and another belly. Then you make it five times, do it many times, ten times trace over, ten times or more, many times, trace it over and when you are tired of tracing, then you try to make it, on your own, below.

The content of this episode concerns the explanation given by the teacher to a child (Eri), who does not know how to write the number 3.

The teacher uses different strategies:

1. She identifies the actions that the child must do: "place your pencil...."
2. She describes the movements: "from the little dot, you will start to draw...."
3. She suggests a repetition of the action in case of failure: "as many times as you need...."

Initially the teacher gives a set of orders indicating actions that the child must do to execute the task. During the realization of the task, when the pupil does not succeed, the teacher continues to explain, but uses another set of strategies:

4. She emphasizes the visual aspect of the symbol 3.
5. She uses the metaphor "belly" to describe the curves of the 3.
6. She returns to strategies 2, 3, 4 and then suggests that Eri repeat the task ("trace the number 3 over again"), giving as a limit the "tiredness" of tracing, a condition that will allow the pupil to write the number 3 without the visual support of the outlining.

In this second instance, we observe the insistence to *trace over* the number 3. To trace over means to outline exactly in the same way as the teacher has done in the model. In fact, instead of stimulating the mental representation of the 3, she is emphasizing the mechanical way to produce it, which is typical of the traditional approach to teaching.

Episode 247

1. *Eri:* Aunt, come and see....
2. *Teacher:* Oh yes, you are doing better now.... At least you are becoming able to do the turn [curve], right Eri?
3. *Eri:* ...
4. *Teacher:* Look Eri, let me tell you, it isn't the right line, no, Eri, look, wait, look, when it is here on the little dot it moves to here and moves to here, all right? You should do two "little bellies," OK? It could be one pregnant mother.... Doesn't one pregnant mother have one belly?
5. *Eri:* Yes....
6. *Teacher:* You must do another pregnant mother, another belly.

In this episode, the teacher first comments on the progress of the pupil in terms of doing the turn (2). The pupil does not react (3), and then she says exactly the opposite of the previous statement (4). She further repeats the same kind of statement that she has given in episode 174, using a similar metaphor.

Example 2: The constructivist approach

Procedure: The teacher and the students are sitting in a circle on the floor. The teacher then places two sets of paper slips on the floor, yellow ones representing units and red ones representing tens. She calls on a pupil and asks the following:

Episode 84

1. *Teacher:* I will ask you for the money and you will pay me, OK? Twenty-two cruzados [a Brazilian monetary unit], what will you do with the slips? (*Pointing to the slips placed on the floor*)

2. *Mar.:* (*Picks two red slips and two yellows*)

3. *Teacher:* How many cruzados are there here? (*Points to the red slips picked by the child*)

4. *Mar.:* 2.

5. *Teacher:* 2. How much are these worth?

6. *Mar.:* 10.

7. *Teacher:* If there are 2, each one has a value of 10, how many cruzados are there?

8. *Mar.:* [Unintelligible]

9. *Teacher:* Mm?

10. *Mar.:* ...

11. *Teacher:* Pay attention, count on.

12. *Mar.:* (*Counts the slips*)

13. *Teacher:* Mm?

14. *Mar.:* [Unintelligible]

15. *Teacher:* So, how many cruzados are there here?

16. *Mar.:* 20.

17. *Teacher:* 20 and the 2?

18. *Mar.:* (*Counts the slips*) 22.

19. *Teacher:* Very good! Let's count to see if there are 22. Here we go! (*Points to the slips*)

20. *All:* 10, 20.

For this episode, in which the pattern of communication between the teacher and the student is based on questioning and answering, we focused on some misunderstandings created during the verbal interaction.

The teacher starts by asking Mar (1) to pick the slips corresponding to twenty-two cruzados, which he does correctly. Then the teacher asks about the quantity of cruzados placed on the floor, pointing to the red slips. The pupil (4) gives an answer corresponding to the number of slips. We can observe a "crossed meaning" event corresponding to these different meanings for the same situation. The teacher asks in terms of money what the value of each slip is, whereas the pupil focuses on the number of slips. Immediately after, the teacher repeats the answer and adds a clarifying question (5). She is trying to repair the meaning of the question to get the answer (6), which is now adequate.

This crossed meaning could correspond to the "perspectival conflicts" (Linell 1995) that frequently occur in a dialogue; however, we want to point out the ability to reconstruct meaning and sensibility that is demonstrated by the teacher in handling the situation in this episode.

Another situation of misunderstanding can occur when the question is very complex, as in (7). Here are three elements in the same question: the number of slips, the value of each one, and the operation to be done. In (8), (10), (12), and (14), the momento during which the pupil is trying to figure out the correct answer—interrupted by the interventions of the teacher in (9), (11), (13)—can be perceived. In (10) is a partial answer that will be completed after the teacher's next question.

Episode 116

1. *Teacher:* Now, And. Take only fifty cruzados.
2. *And.:* (*Picks five red slips*)
3. *Teacher:* Now count them!
4. *And.:* 10, 20, 30, 40, 50 (*in a faint voice*).
5. *Teacher:* Count a little bit louder. Can you hear her?
6. *All:* No!
7. *Teacher:* Then let's ask her to count louder.
8. *And.:* 10, 20, 30, 40, 50.
9. *Teacher:* How many red slips are there?
10. *Some pupils:* 50.
11. *Teacher:* How many reds?
12. *And.:* 5.
13. *One pupil:* 10.
14. *Teacher:* 5. How much value does one red slip have?

15. *And.:*   10.
16. *Teacher:* 10. And how many slips are there?
17. *And.:*   [Unintelligible]
18. *Teacher:* Mm?
19. *And.:*   5.
20. *Teacher:* 5. If we ... if the value of each one is 10, how will we count? 10....
21. *And.:*   10, 20, 30, 40, 50.
22. *Teacher:* 50. Very good! Now Fla.

The content of this episode is also concerned with counting slips: *quantity* and *value*. The pupil And. identifies five red slips corresponding to fifty cruzados (2). But when the teacher tells her to count (3), she is not so sure and counts in a low voice. She seems uncertain whether the teacher is waiting for an answer corresponding to the number of slips or to the value of the slips. Really, the teacher's question refers only to performing the action; it is not clear if the child has to count the number or the value of the slips.

After the pupil had counted in tens, taking into account the value, (4) and (8), the teacher asks her about the quantity of slips. What do we observe? Some students answer the previous question. It seems that they have not registered the new question.

How does the teacher deal with these types of answers? She repeats the previous question (11). Pupil And. gives the expected answer, but one pupil gives an answer that is still based on the value of the red slips (13). The teacher does not pay any attention to this loose answer and continues the dialogue with And. She repeats And.'s answer and asks about the value of the red slips (14). And. gives the expected answer (15). The teacher returns to the question about the quantity of slips (16), and in the end And. returns to counting on the basis of the value of slips.

In this episode, the dialogue is conducted by the teacher, who introduces all the questions. The pupil is given the role of "répondeur quasi automatique" (quasi answering machine). Actually it was the teacher who had set the agenda by defining the content of the discussion, orchestrating the dialogues, and establishing the criteria of relevance and appropriateness of the contributions that the pupils could give (Edwards and Mercer 1987, p. 131). In terms of content, a going back and forth occurs between the concepts of value and quantity, which has not always been perceived by the pupils.

DISCUSSION

The patterns from the first classroom, episodes 093, 098, 134, 174, can easily be found in the sequence of the whole protocol and also in others (research in progress). They show the sterility of an approach that focuses on results and ignores the ways that children can use to arrive at their results. The repetition of models, drills, and exercises can provoke a mechanical acquisition of knowledge.

Meanwhile, we find it interesting to observe how the same teacher, when handling a more difficult student, uses a metaphor as a means of explaining the task that, in this particular situation, conjures up a mental representation of the form asked for. Pimm (1992) clearly mentions how meaning also comes from images as well as how language can be used creatively. The simple form of the symbol representing the number 3 had not in itself been so evident to this child. In fact, he needed the support of an image situated in an everyday context to help him with such symbolization.

Concerning the episodes in the second classroom, we can observe a totally different pattern of communication that is based on questioning and answering, although in a quasi-mechanical way. The repetition of the same kind of question has been observed with regularity. This constant repetition has created a sort of misunderstanding, a miscommunication. In fact, the studies of Hundeide (1992) have shown that asking the same question twice can provoke a new interpretative premise. "If we take everyday conversational practice as a point of departure, would it not be reasonable to assume that a new reply is expected to the second question, when a person asks the same question twice? Repeating a question is, after all, the way authoritarian schoolmasters and other adult 'care persons' use to convey their dissatisfaction with a first reply" (Hundeide 1992). Grossen and Perret-Clermont (1984) mention that in test situations, when the adult is in control of the event, the child tends to adopt a submissive pupil role, and tries to follow and adjust his or her thinking to the adult requests.

FINAL REMARKS

Traditional and constructivist are two approaches to teaching that coexist in teacher-student communication. Even constructivist teachers still use the old patterns of repetition simultaneously with more comprehensive strategies. This situation shows how difficult it is to change patterns of communication. One reason might be related to the content of educational reforms. In general, they are based on only curricula changes. Communication patterns, which are far from being just an accessory to the process of learning, are neglected, especially since this particular aspect of teaching and learning has always been taken for granted. Rather than go through a wave of well-intentioned reforms, based on beliefs about what should happen, we suggest using our modest cultural accounts of what actually happens in schools and how mathematical knowledge is actualized in the everyday life of classrooms. Seeger (chapter 4 in this volume) points to the need "to better understand what is the rationality of the prevailing classroom discourse that underlies the recurrent patterns." In fact, the instructional discourse is also an expression of the learning concept elaborated by teachers during the training years as well as in actual classroom experience.

More concretely, we think that analyses of such episodes of communication between teacher and students can be very useful in didactics courses and also in the in-service training of teachers in different ways:

- To offer possibilities to relate the experience of the actual situation of protocols with their own previous experience as a teacher
- To identify precise strategies that can be removed or improved to get a more effective and comfortable communication with students
- To analyze critically their own present practice as well as to be able to conceive of alternatives

REFERENCES

Bishop, A. J., and G. Pompeu Jr. "Influences of an Ethnomathematical Approach on Teacher Attitudes to Mathematics Education." In *Proceedings of the Fifteenth Conference of International Group for the Psychology of Mathematics Education*, pp. 136–43. Assisi, Italy: International Group for the Psychology of Mathematics Education, 1991.

Carraher, T. N., ed. *Aprender pensando: Contribuicoes da psicologia cognitiva para a educaçao*. Petropolis, RJ.: Vozes, 1989.

Carraher, T. N., D. W. Carraher, and A. D Schliemann. "Mathematics in the Streets and in Schools." *British Journal of Developmental Psychology* 3 (1985): 21–29.

———. *Na vida dez, na escola zero*. Sao Paulo: Cortez, 1988.

Coulthard, M. *An Introduction to Discourse Analysis*. London: Longman, 1990.

Edwards, D., and N. Mercer. *Common Knowledge: The Development of Understanding in the Classroom*. London: Methuen, 1987.

Garfinkel, H. *Studies in Ethnomethodology*. Englewood Cliffs, N.J.: Prentice-Hall, 1967.

Garvey, C. *Children's Talk*. Oxford: Fontana, 1984.

Grossen, M., and A. N. Perret-Clermont. "Some Elements of a Social Psychology of Operational Development of the Child." *Quarterly Newsletter of Laboratory of Comparative Human Cognition* 6 (1984): 51–57.

Hundeide, K. "The Tacit Background of Children's Judgments." In *Culture, Communication, and Cognition: Vygotskian Perspectives*, edited by J. V. Wertsch, pp. 306–22. Cambridge: Cambridge University Press, 1988.

———. "The Message Structure of Some Piagetian Experiments." In *The Dialogical Alternative: Towards a Theory of Language and Mind*, edited by A. H. Wold, pp. 139–56. Oslo: Scandinavian University Press, 1992.

Inhelder, B., H. Sinclair, and M. Bovet. *Apprentissage et structures de la connaissance*. Paris: Presses Universitaires de France, 1974.

Lautrey, J. *Classe sociale, milieu familiale, intelligence*. Paris: Presses Universitaires de France, 1984.

Linell, P. "Troubles with Mutualities: Towards a Dialogical Theory of Misunderstanding and Miscommunication." In *Mutualities in Dialogue*, edited by I. Marková, C. F. Graumann, and K. Foppa. Cambridge: Cambridge University Press 1995.

Linell, P., and I. Marková. "Acts in Discourse: From Monological Speech Acts to Dialogical Inter-acts." *Journal for the Theory of Social Behaviour* 23 (1992): 173–95.

Marková, I., and K. Foppa. *The Dynamics of Dialogue*. New York: Harvester and Wheatsheaf, 1990.

Patto, M. H. S. *A producao do fracasso escolar*. Sao Paulo: T. A. Queiroz, 1991.

Perret-Clermont, A. N. *A construcao da inteligência pela interacao social*. Lisbon: Sociocultur, 1978.

——— "Transmitting Knowledge: Implicit Negotiations in the Teacher-Student Relationship." In *Effective and Responsible Teaching: The New Synthesis*, edited by F. K. Oser, A. Dick, and J. L. Patry, pp. 329–41. San Francisco: Jossey Bass, 1992.

Perret-Clermont, A. N., J. F. Perret, and N. Bell. "The Social Construction of Meaning and Cognitive Activity in Elementary School Children." In *Perspectives on Socially Shared Cognition*, edited by L. B. Resnick, J. M. Levine, and S. D. Teasley, pp. 41–62. Washington, D.C.: American Psychological Association, 1991.

Pimm, D. "Metaphoric and Metonymic Discourse in Mathematics Classrooms." Milton Keynes, England: The Open University, Centre for Mathematics Education, 1992,

Rommetveit, R. *On Message Structure: A Framework for the Study of Language and Communication*. New York: Wiley, 1974.

———. "Outlines of a Dialogically Based Social-Cognitive Approach to Human Cognition and Communication." In *The Dialogical Alternative: Towards a Theory of Language and Mind*, edited by A. H. Wold. Oslo: Scandinavian University Press, 1992.

Taylor, T. J., and D. Cameron. *Analysing Conversation: Rules and Units in the Structures of Talk*. Oxford: Pergamon Press, 1987.

Vygotsky, L. S. *Mind and Society: The Development of Higher Psychological Processes*. Cambridge: Harvard University Press, 1978.

———. *Thought and Language*. Cambridge: MIT Press, 1988.

Wold, A. H. *The Dialogical Alternative: Towards a Theory of Language and Mind*. Oslo: Scandinavian University Press, 1992.

9

Alternative Patterns of Communication in Mathematics Classes: Funneling or Focusing?

Terry Wood

Purdue University, West Lafayette, Indiana

I T HAS been well acknowledged that the manner in which teachers and students interact reflects not only the routines for harmonious functioning in the class but also the nature of the learning opportunities that may occur for children. Lessons in mathematics classrooms can be characterized by interaction patterns and ways of communicating that, to the observer, reveal the different views about teaching and learning mathematics that are held by the participants. Thus, the dialogue that is found in classrooms marks the "stance of the speaker towards the event being represented, towards the occasion of utterance, and towards the manner in which the speaker expects the listener to view the world and use his mind" (Bruner 1986, p. 125).

For example, Voigt (1985) and Steinbring (1993) both give illustrations of classroom communication that appear to be situations in which the teachers' goal is to allow students to express their personally constructed solutions to mathematical problems. However, on closer examination, it is seen that what the teachers do is wait until a student gives a solution that is representative of the method they intend the students to use. At this point, they step in and highlight the solution and by so doing emphasize that *this* is the method for students to know.

Some aspects of the research reported in this paper were supported by the National Science Foundation (RED-92544939). All opinions expressed are solely those of the author.

Perhaps not realizing it, they are at the same time communicating to the student that what is most important in mathematics is finding the predetermined solution that they, the teachers, have in mind.

A different pattern of interaction identified in classrooms is described by Wood (1994). In these classrooms, students are encouraged to express their thinking but, although a variety of solutions are accepted and valued, the teacher may highlight a solution that he or she finds is an interesting way to think about the problem. On the surface, this alternative interaction may seem similar to the one previously described. However, in this situation, attention is given to the method, but not because it reflects the teacher's predetermined solution. Instead, the strategy is emphasized to help students notice an idea. Having accomplished this goal, the teacher continues the dialogue by again asking for other solutions to the problem.

In the light of these examples, it is clear that analyzing the differences in the ways in which teachers and students communicate allows us to examine in greater detail the process of teaching and learning mathematics within the complexity of the classroom. To this end, the purpose of this chapter is to extend the existing analyses of communication in classrooms by considering the interplay between the ways of communicating and the opportunities that occur for students to learn mathematics with understanding.

THE ROLE OF COMMUNICATION IN STUDENT'S LEARNING

Wertsch and Toma (1995) characterize communication in classrooms as being either univocal or dialogic. Drawing on the work of Lotman, who claimed that all "texts" are characterized as having two functions, a univocal function and a dialogic function, Wertsch and Toma seek to explain the role of discourse in students' learning. On the one hand, the univocal function of text is described as emphasizing the transmission aspect of communication; the text is seen as a passive link in conveying information from a sender to a receiver. On the other hand, text is also described as having a dialogic function in which it serves as a thinking device to act as a generator of new meanings for the receiver. These constructs become very useful in trying to understand classroom talk in which the function can be either to transfer mathematical knowledge to students or to act as a means for enabling students to generate new meanings for themselves. In addition, the proposed changes currently advocated for reform in mathematics education support a more dialogic view of communication (NCTM 1989, 1991).

TRADITIONAL PATTERNS OF COMMUNICATION IN MATHEMATICS CLASSROOMS

The regularity and structure of classroom discourse have been previously considered by Sinclair and Coulthard (1975) and Mehan (1979). These authors describe the typical communication pattern found in classrooms as the recitation pattern. This discourse has the well-known IRE sequence, in which the teacher initiates a question to which the student responds; this response is then evaluated for correctness by the teacher. Furthermore, these studies show that lessons are structured in predictable ways that enable the participants to know how to speak and about what to speak. For the most part, however, this tripartite sequence can be viewed as attempts by the teacher to communicate to students in a way that reflects the univocal function of text described by Wertsch and Toma. Thus, learning mathematics in school has remained largely unchanged over the past twenty five years. Meserve and Suydam (1992) describe mathematics instruction as showing the children what to do and giving them practice in it. Likewise it has been argued that classrooms in which communication occurs in a univocal manner may not support students' learning of mathematics with meaning.

ALTERNATIVE PATTERNS OF COMMUNICATION IN MATHEMATICS CLASS

Results from both quantitative and qualitative research indicate that students learn mathematics with greater understanding in classrooms in which they are allowed to explore, investigate, reason, and communicate about their ideas (e.g., Hiebert and Wearne [1993]; Wood and Sellers [1996]). Thus, one major difference between traditional mathematics classes and these alternative classes lies in the nature of the *social norms* (Yackel, Cobb, and Wood 1991) that are interactively constituted among the participants and that underlie the different patterns of interaction that evolve.

The smooth functioning of the classroom, as seen from the observer's perspective, evolves from the mutual construction of expectations and obligations between teachers and students. The social norms underlie the patterns and routines that become established in the classroom and that enable the students and teacher to interact harmoniously. The patterns become the *hidden regularities* that guide the actions of the participants in the classrooms. As such, they become the taken-for-granted ways of interacting that constitute the culture of the classroom. In attempting to deepen the analysis of the nature of communicating in classrooms and the subsequent opportunities for learning that are created, it is necessary to consider the underlying social norms that are constituted. The *interlocking networks of obligations and expectations* (cf. Goffman [1974, p. 346]) that exist for both the teacher and students influence the regularities by which students and teacher interact and create opportunities for communication to occur between the participants. These interaction patterns serve to constrain or enhance children's opportunities to actively construct mathematical meaning. Opportunities for children to reflect on their own understanding and reasoning about mathematics occur more frequently in situations in which they are allowed to express and clarify their mathematical thinking to others. In addition, as listeners in these settings, children have opportunities to reflect on the mathematical reasoning and ideas of others.

In deepening the analysis of children's learning in mathematics classrooms, it is important to consider both the form of the discourse (knowing how to talk) and the content (knowing what to say) as a means of describing how the participants communicate with one another. The work of Bauersfeld (1988), Bauersfeld, Krummheuer, and Voigt (1988), and Voigt (1985) extends previous analyses of classroom communication to include both form and content in their investigations of the interaction that occurs in traditional mathematics classes. One frequent way in which teachers and students communicate is described by Bauersfeld (1980) as the funnel pattern.

Funnel Pattern

The following episode is presented to illustrate the nature of the interaction that is referred to as the funnel pattern. In this example, the teacher has just asked Jim to give the answer to 9 + 7.

Jim: 14.

Teacher: OK. 7 plus 7 equals 14. 8 plus 7 is just adding one more to 14, which makes ___? (*voice slightly rising*).

Jim: 15.

Teacher: And 9 is one more than 8. So 15 plus one more is ___?

Jim: 16.

The funnel pattern began when Jim gave an incorrect answer. In the traditional recitation pattern of interaction, a teacher would simply tell the student that his or her answer was wrong and seek another child to give the correct response. But in this example, the teacher attempts to resolve the situation in a different way. The teacher, at first glance, seems to accept Jim's incorrect answer and to intend to let him rethink his response. However, later it can be seen that this is not the situation. Instead, the incorrect answer is a starting point from which the teacher leads Jim through a series of explicit questions until he provides the correct answer. In this example, from the observer's perspective, it appears that the teacher's intention is for the student to see that any unknown basic addition fact can be determined by using the strategy of "the double plus one."

Current recommendations suggest to teachers that it is important for students to develop these thinking strategies as a way to learn the so-called basic facts. A number of research studies have shown that by using thinking strategies, children are better able to derive unknown facts (Rathmell 1978; Thornton 1978; Treffers 1991). Finding basic facts in this manner enables children to develop an understanding of numbers and relationships. Therefore, from an instructional perspective, teaching children to know and use these strategies will in fact help them learn the basic facts.

Superficially, this does appear to be the outcome of the exchange. However, a closer examination of the dialogue reveals that the student's role in the exchange, mathematically, consists of nothing more than adding one more to the sum given by the teacher. Furthermore, to participate in this exchange, the student does not need to think about the relationships that exist between numbers to give the correct response to the teacher's questions. Instead, all that is necessary for the student to be effective is to fill in the correct missing number word to complete the blank in the teacher's question. To all intents and purposes, the student may have interpreted the task as simply "add one to the number given by the teacher." If so, the opportunity for the student to learn about using thinking strategies to derive unknown facts has been missed.

The teacher, in contrast, is the one who actually engages in the cognitive activity of using these strategies as she creates the series of related number sentences that are intended to help Jim generate the correct answer for 9 plus 7. Moreover, despite the teacher's well-intentioned questioning to guide the student through the procedure, the student actually had very little opportunity to engage in meaningful mathematical activity. The communication and interaction are, in Wertsch and Toma's term, univocal, and the student's thinking is focused on trying to figure out the response the teacher wants instead of thinking mathematically for himself. Thus, although the teacher may intend that the child use strategies and learn about the relationships between numbers, the student needs to know only how to respond to the surface linguistic patterns to derive the correct answers. Lundgren (1977) claims that a pattern of communication such as this "gives the illusion that learning is actually occurring" (p. 202), although in fact it is not.

Focusing Pattern

Another alternative to the recitation pattern found in reform-oriented mathematics instruction is to design situations in which students learn as they participate more equally in the dialogue. A high level of interaction between the teacher and students creates opportunities for children to reflect on their own thinking and on the reasoning of others. The example that follows is drawn from a classroom in which children's mathematical thinking and meanings are the central emphasis of the instructional program.

The teacher in this instance creates situations that allow children to explain and give reasons for their mathematical ideas as they communicate to others in the classroom. Moreover, she expects students to be thinking about mathematics and to be discussing their ideas with others. The children, for their part, expect that the teacher will then accept their thinking and respect their ideas. The interplay between these expectations and obligations for one another's behavior is reflected in the nature of the communication that occurs. As some background to the example presented, the children are in the process of discussing their self-generated methods for solving two-digit subtraction problems with regrouping. It should be understood that in this class, the traditional standard algorithms are not taught directly and the children are encouraged to invent various ways to solve multidigit problems (Labinowicz 1985). The problem the students are solving is 66 − 28 = ___. The teacher has asked John to tell the class his solution to the problem.

John: (*He is writing on the overhead projector.*) We put the 28 under the 66. (*As he talks, he writes 66 − 28 in a vertical format.*) And we took away ... we ... I took ... the 6 and 8 off. And we said there was 60 and 20 there. (*He puts his finger on the 60 and then on 40.*) And if you take away 20 from 60, it's 40. (*He holds up his fist.*) And you still have to take away 8. So we took ... there's 46 left over. If you take that 6 back, and take away that 6 (*points to the 6 in 46*) and that's ... um ... back to 40 and you still have to take away 2, so 39 (*he holds up a finger*) then 38. (*He writes 38.*)

Listening, the teacher realizes that John's solution is new for him and that it is not one that has been given previously in the class. Prior to this, many students in this class have solved two-digit addition problems by splitting the numbers into tens and ones, adding the tens, and then operating with the ones. This collection strategy (Cobb and Wheatley 1988), although effective and efficient in addition, leads to new conflicts when students encounter a problem like 66 – 28. The problem arises not with the tens, but with the ones. Typically, some students attempt to resolve the situation by reframing the problem to 8 – 6. However, finding this strategy unsuccessful, they are confronted with a situation that they are then unable to resolve immediately.

In listening to John, the teacher recognizes that he has invented an effective solution and that his strategy may help others. Thus, rather than stepping in and repeating his idea, she asks the class if anyone has a question. One boy says that he does not understand, and the teacher then asks John to reexplain.

John: We put the 66 under the 28. Then we took off the 6 and the 8 and, if you take away 20 plus 60, it's 40. And if you put the 6 back on and the 8, we have 46. Then we take away ... we still have to take away that 8. Then you take away that 6, now you have 40 back and you still have to take away 2.

Elisabeth: But, but why did you take the 6 and the 8 off?

John: It was more easier.

Teacher: (*Looking around at the class and deciding that the others still may not understand what he did*) OK, could you write down beside it what you did? Maybe that would help us see it. Instead of 66 minus 28, what did you do?

John: 60 take away 20 equals (*he writes 60 – 20 vertically, and looks at the teacher*).

Teacher: Would you write what you get? (*He writes 40 under 60 – 20.*). OK, what did you do next?

Having asked John to explain his thinking by using written symbol notation, she enables him to clarify his thinking to the students. She knows that his comment "It was more easier" does not fully answer Elizabeth's question. Yet, she also realizes that the strategy of subtracting 20 from 60 is one that is familiar to most of the students. Once John writes this, she asks, "OK, what did you do next?" It is this question that serves to focus the joint attention of the class on the remainder of his solution that may be helpful to others.

John: Then we put the 6 back on. Then it equaled 46. (*He writes + 6 next to the 40.*) And you still have to take away 8, so you have 40 back. And ... um ... if you take away 2—you have to take away 2 more, so we got 38.

Teacher: (to the class). Make sense? (*Pause*) Do you understand what he said about his part? He said (*coming to the front of the class to use the overhead projector*) I have, let's put this 46 up here (*she writes 46 at the top*). That's what he has and then he said I've got to go back to 40. Okay, why did you go back to 40?

From the observer's perspective, the teacher realizes that some of John's explanation may not be understood by the others. At this point, she decides to step in and summarize the aspects of the solution that she feels most of the children understand. Then, intending to draw attention to the point at which she feels that John's reasoning has been difficult for the others to follow, she asks, "Why did you go back to 40?" She rephrases John's original statement, "So you have 40 back," as an action he has performed in an attempt to focus attention on the discriminating aspect of his solution strategy so he will provide a rationale for the rest of the class.

John: 'Cause we took away that 6, 'cause you have to take away 8 and you still have to take away 2 more.

Teacher: You understand how he did that?

Class: Yeah.

Teacher: (*Long pause*) Very interesting way to do that. Thank you.

In this episode, the teacher's focusing question serves as an attempt to orient the discussion to the aspect of John's solution that is distinctive by not allowing the focus of the children's attention to fade or change or be interrupted. From the observer's perspective, the focusing pattern differs significantly from the funneling pattern as a way of communicating. Although the teacher intervenes to ask a question that is directed toward a specific response, the goal in doing so reflects a different stance toward the learning of mathematics. Rather than attempt to guide John in such a way as to funnel his thinking, she instead turns the control of the conversation back to him. In this way, he is responsible for reexplaining his thinking to the others. In addition, the focusing question of the teacher directs the attention of the students who are listening to the salient features of the solution. But then, by stepping out of the conversation, the teacher affords students the opportunity to make sense of the strategy for themselves. The teacher, in this situation in which a unique solution has been given, necessarily tries to anticipate what the other students might not understand and asks clarifying questions to keep attention focused on the discriminating aspects of the solution.

At the end of the episode, she does not intend that all students will now use John's method, but only that they try to follow his reasoning and make sense of it. John's solution represents just one of the many ways in which children solve subtraction problems in this classroom. In this way, the teacher communicates to her students that although their own personal mathematical ideas are of value, the ideas of other students are of equal importance.

Interplay between Social Norms and Interaction Patterns

The regularities in behavior that are held by members of the classroom community are built over the course of the school year between the teacher and students. An interlocking system of obligations and expectations, established by both the teacher and the students and underlying the manner in which members of the classroom interact, forms the smooth functioning of the class. The interplay of social norms with patterns of interaction creates a different culture for learning mathematics in classrooms.

The alternative communication pattern, described as funneling, reflects certain beliefs about the nature of mathematics and the relationship between teacher and students. The funnel pattern, although it is an alternative to the traditional pattern of directly telling a student the correct procedure or answer, still conveys a view that the mathematics to be learned rests solely within the authority of the teacher. Although the teacher's intention is to help students learn by guiding them until they obtain a correct answer, from the students' perspective, learning mathematics still involves determining a set of procedures that the teacher already knows and that it is their obligation to learn.

The alternative pattern, described as focusing, exists in a class in which the teacher expects the students to think about mathematics, to figure out things for themselves, and to discuss their ideas with others. In this setting, the teacher expects the children to communicate their ways of solving problems to one another. However, from the students' point of view, a certain amount of risk is involved in a situation in which their thinking is to be made available for public evaluation. In this setting, students are responsible for explaining their solution methods to their classmates, who are expected to ask questions for clarification and justification. The teacher and students have created a community in which students are actively involved in the process of communicating about mathematics. The teacher, to meet her obligations to her students, has developed ways of interacting with her students that enable them to give explanations and reasons for their thinking. To keep a balance between guiding students in their understanding and allowing them to engage in their own constructive activities, the teacher has developed ways of communicating that serve to focus students' joint attention on aspects of a solution not yet fully understood but that then leave to them the responsibility of solving the problem. In this way the teacher has conveyed to the students that what counts as mathematics in her class are the meanings and understandings that the children have constructed for themselves.

ACKNOWLEDGMENTS

Several of the ideas central to this paper were elaborated in the course of previous discussions with Heinrich Bauersfeld, Götz Krummheuer, and Jörg Voigt at the University of Bielefeld. The responsibility of interpretation is the author's and may not reflect the views of the others.

The notion of focusing is the result of wonderfully enjoyable e-mail discussions with David Pimm from The Open University. (*Editors' note:* Compare the focusing pattern with a pattern of language use in the classroom identified by Edwards and termed "focus and filter" [cited in Edwards, D., and N. Mercer. *Common Knowledge: The Development of Understanding in the Classroom*, 2nd ed. London: Routledge, 1993].)

REFERENCES

Bauersfeld, H. "Hidden Dimensions in the So-Called Reality of a Mathematics Classroom." *Educational Studies in Mathematics* 11 (1980): 23–41.

———. "Interaction, Construction, and Knowledge: Alternative Perspectives for Mathematics Education." In *Effective Mathematics Teaching*, edited by T. Cooney and D. Grouws, pp. 27–46. Reston, Va.: National Council of Teachers of Mathematics, 1988.

Bauersfeld, H., G. Krummheuer, and J. Voigt. "Interactional Theory of Learning and Teaching Mathematics and Related Microethnical Studies." In *Foundations and Methodology of the Discipline of Mathematics Education*, edited by H. G. Steiner and A. Vermandel, pp. 174–88. Antwerp: Psychology of Mathematics Education, 1988.

Bruner, J. *Actual Minds, Possible Worlds*. Cambridge: Harvard University Press, 1986.

Cobb, P., and G. Wheatley. "Children's Initial Understandings of Ten." *Focus on Learning Problems in Mathematics* 10 (1988): 1–28.

Goffman, E. *Frame Analysis: An Essay on the Organization of Experience*. Cambridge: Harvard University Press, 1974.

Hiebert, J., and D. Wearne. "Instructional Tasks, Classroom Discourse, and Students' Learning in Second-grade Arithmetic." *American Educational Research Journal* 30 (1993): 393–425.

Labinowicz, E. *Learning from Children: New Beginnings for Teaching Numerical Thinking*. Menlo Park, Calif.: Addison-Wesley, 1985.

Lundgren, U. P. *Model Analysis of Pedagogical Processes*. Stockholm: Stockholm Institute of Education, Department of Educational Research, 1977.

Mehan, H. *Learning Lessons: Social Organization in the Classroom*. Cambridge: Harvard University Press, 1979.

Meserve, B., and M. Suydam. "Mathematics Education in the United States." In *Studies in Mathematics Education: Moving into the Twenty-first Century*, edited by R. Morris and A. Manmohan. Paris: UNESCO, 1992.

National Council of Teachers of Mathematics (NCTM). *Curriculum and Evaluation Standards for School Mathematics*. Reston, Va.: National Council of Teachers of Mathematics, 1000.

———. *Professional Standards for Teaching Mathematics*. Reston, Va.: National Council of Teachers of Mathematics, 1991.

Rathmell, E. C. "Using Thinking Strategies to Teach the Basic Facts." In *Developing Computational Skills*, edited by M. Suydam and R. Reys, pp. 13–38. Reston, Va.: National Council of Teachers of Mathematics, 1978.

Sinclair, J., and R. Coulthard. *Towards an Analysis of Discourse: The English Used by Teachers and Pupils*. Oxford: Oxford University Press, 1975.

Steinbring, H. "Problems in the Development of Mathematical Knowledge in the Classroom: The Case of a Calculus Lesson." *For the Learning of Mathematics* 13 (1993): 37–50.

Thornton, C. "Emphasizing Thinking Strategies in Basic Fact Instruction." *Journal for Research in Mathematics Education* 9 (1978): 214–27.

Treffers, A. "Meeting Innumeracy at Primary School." Educational Studies in Mathematics 22 (1991): 333–52.

Voigt, J. "Patterns and Routines in Classroom Interaction." *Recherches en didactique des mathématiques* 6 (1985): 69–118.

Wertsch, J. V., and C. Toma. "Discourse and Learning in the Classroom: A Sociocultural Approach." In *Constructivism in Education*, edited by L. P. Steffe and J. Gale, pp. 159–74. Hillsdale, N.J.: Lawrence Erlbaum Associates, 1995.

Wood, T. "Patterns of Interaction and the Culture of Mathematics Classrooms." In *The Culture of the Mathematics Classroom*, edited by S. Lerman, pp. 149–68. Dordrecht, Netherlands: Kluwer Academic Publishers, 1994.

Wood, T., and P. Sellers. "Assessment of a Problem-Centered Mathematics Program: Third Grade." *Journal for Research in Mathematics Education* 27 (1996): 337–53.

Yackel, E., P. Cobb, and T. Wood.. "Small-group Interactions as a Source of Learning Opportunities in Second-grade Mathematics." *Journal for Research in Mathematics Education* 22 (1991): 390–408.

10

Students Communicating in Small Groups: Making Sense of Data in Graphical Form

Frances R. Curcio
New York University

Alice F. Artzt
Queens College of the City University of New York

A RECOGNITION of the limitations of isolated, static drill and practice and the advantages of collaborative, dynamic learning has redirected attention to the nature of the mathematical language employed by students and the communication that occurs among students in the mathematics class (NCTM 1989, 1991). A study that examined the problem-solving behaviors within the context of small groups (Artzt and Armour-Thomas 1992) suggests that the continuous interplay of cognitive and metacognitive behaviors that occurs among members of successful problem-solving groups mirrors the thoughts and behaviors of expert problem solvers working alone (Schoenfeld 1987). Students working in small groups have the opportunity to communicate about mathematics as they discuss and develop problem-solving

---

This chapter is based on a paper titled "The Effects of Small Group Interactions on Graph Comprehension of Fifth Graders," presented at the Seventh International Congress on Mathematical Education, Quebec, 1992.

Thanks are extended to Professor Eleanor Armour-Thomas, who assisted in the design of the study and the analysis of the data reported herein.

strategies and deal with resolving misconceptions (Stacey and Gooding, chapter 11 in this volume).

One type of problem-solving task that elicits both cognitive and metacognitive activity on the part of the reader involves the analysis of, and extraction of meaning from, visual displays of data (Curcio 1987). Middle school students' difficulties in interpreting data (Kouba et al. 1988) may be related to difficulties they have in actively monitoring and subsequently regulating cognitive processes (Schoenfeld 1987) they engage in while extracting meaning from a visual display.

The exploratory study described in this chapter is based on the work of Artzt and Armour-Thomas (1992), who developed a framework for analyzing the interactions among the members of small groups. With a graphing activity as the task for the small group, the cognitive-metacognitive framework from Artzt and Armour-Thomas was employed to examine the effects of small-group communication on data interpretation by fifth graders. In this chapter, the problem-solving framework is presented, the graph task is described, the framework is applied to the graph task, examples of student interactions are analyzed, and implications for instruction are presented.

THE PROBLEM-SOLVING FRAMEWORK

The instrument used to evaluate the nature of the communication of students during small-group work is based on a problem-solving perspective. The categories of problem-solving behavior within a small group are defined as read, understand, explore, analyze, plan, implement, verify, watch, and listen. To locate the instances of monitoring and regulating the problem-solving enterprise, each behavior is categorized as either cognitive or metacognitive. The working distinction between cognition and metacognition is that cognition is involved in doing, whereas metacognition is involved in choosing and planning what to do and monitoring and regulating what is being done. Metacognitive behaviors can be exhibited by statements made *about the problem* or statements made *about the problem-solving process.* Cognitive behaviors can be exhibited by verbal or nonverbal actions that indicate the actual processing of information. (See table 1 in Artzt and Armour-Thomas [1992] for an outline of the categorization of problem-solving behaviors.) A justification for the cognitive and metacognitive assignments follows.

Analyzing and planning are, by their nature, predominantly metacognitive behaviors. Any statements revealing that one is trying to simplify or reformulate a problem are necessarily statements made *about* the problem or *about* the problem-solving process. Likewise, planning would be indicated by statements made about how to proceed in the problem-solving process. Understanding the problem is categorized as predominantly metacognitive because this category is indicated only when students make comments that reflect attempts to clarify the meaning of a problem. That is, when a student makes a comment regarding the meaning of a problem, he or she is making a comment *about* the problem. Reading behaviors are categorized as predominantly cognitive because they exemplify instances of "doing." Exploring, implementing, and verifying behaviors are sometimes categorized as cognitive and sometimes as metacognitive. Exploration at the cognitive level alone often results in disorderly, aimless, and unchecked wanderings. When exploration is guided by the monitoring of either oneself or one's group member, that behavior can be categorized as exploration with monitoring or exploration with metacognition. Such monitoring leads to self- or group-regulation of the exploration process, thereby keeping the exploration controlled and focused. A similar analysis applies for implementation and verification, which can also occur with or without monitoring and regulation. Watching and listening behaviors cannot be categorized as either cognitive or metacognitive because these are not audible behaviors. However, watching and listening are equal partners with speaking in the communication process. (A detailed description of the framework is included in Artzt and Armour-Thomas [1992].)

The Graph Task

Chuck, Dennis, Garin, and Razzie, four fifth graders in a New York City middle school, were given a broken line graph depicting the average time of sunset from June to December (see fig. 10.1). They were asked to draw a picture of what the graph would look like if it were to continue from January through May.

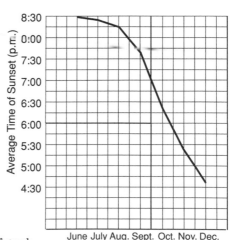

Fig. 10.1. The "average time of sunset" graph

APPLYING THE GRAPH TASK
TO THE FRAMEWORK

Before the problem-solving behaviors of the students are described and an actual protocol example is given, the graphing task with respect to the framework is examined. An outline is presented of several approaches that could be used, many of which the students did use. These approaches are described by possible problem-solving behaviors and their related cognitive levels. (The order is not necessarily the sequence in which they may occur.)

- *Reading the problem (cognitive).* The students read or listen to someone else read the problem.
- *Understanding the problem (metacognitive).* The students understand that a graph must be drawn from January to June. They recognize that three things are needed to complete the task.

  1. A rectangular grid must be drawn.
  2. The grid must have an x-axis containing the months January, February, March, April, and May.
  3. The y-axis must contain the average times of sunset, which they are to figure out.

- *Analyzing the problem (metacognitive).* The students critically examine the elements of the problem. They may analyze the problem in the following ways:

  1. They can look at the patterns from the given data, and they can notice a trend in the change of average time of sunset from one month to the next.
  2. They can think about their personal experiences with sunset hours in the months from January through May.
  3. They can think about what the average time of sunset might be if the pattern were to continue and compare that with their experiences.

- *Planning (metacognitive).* If the students attempt to plan an approach for solving the problem, they may decide to do the following:

  1. Draw a coordinate grid with the months from January through May across the x-axis.
  2. List the average times of sunset on the y-axis (the range of these times differs).

3. Find the difference between each successive average time of sunset and look for a pattern in the differences.
4. Continue the pattern of differences to calculate the average time of sunset in the remaining months by subsequent subtractions.
5. Use the pattern of differences to create a pattern of subsequent additions to find the remaining average times of sunset.
6. Use personal experiences of all group members to get the remaining average times of sunset.

Note that the plans might not be the ones that will lead to a successful solution.

- *Exploring (cognitive and metacognitive).* This problem lends itself to some guessing and testing because the differences in the data, in successive months, of average times of sunset do not conform to a consistent pattern. Although the average time of sunset from June through December does steadily decrease, it does not do so in a discernable pattern. This means that students usually begin estimating what the continuation of the pattern could be. Without monitoring, this exploration might lead to an incorrect continuation of the decrease in average time of sunset. Another unmonitored exploration might lead a student just to continue the line formed by the given graph. When monitored, these types of explorations could lead to students' realizing that a change in direction must be made because their experiences do not support their results.

- *Implementing (cognitive and metacognitive).* When a student has devised a plan for solving the problem, he or she is likely to try to implement the plan. If the student performs this implementation systematically through monitoring and regulation (metacognitive), the student is likely to discover that the plan was good and has led to a reasonable solution or that it was faulty and needs some adjustment. If the implementation is unmonitored (cognitive alone), however, the student may follow through on the implementation leading to an incorrect and unreasonable solution.

- *Verifying (cognitive and metacognitive).* An effective verification requires the student to examine his or her final solution and check that the answer makes sense (metacognitive). That is, are the average times of sunset they have calculated consistent with their personal observations of average times of sunset in the months January through May? If they have tried to continue the pattern in the given data as it is or in the reverse direction, they may try to verify their results by checking their successive subtractions or additions (cognitive).

- *Watching and Listening (uncategorized)*. Students must be willing and able to listen and watch one another for an exchange of ideas to take place.

An Analysis of an Example of Student Communication

To categorize the problem-solving behaviors of the students in a way that would give a global picture of the dynamics of the communication, a coding chart was used. The chart identifies students' problem-solving statements and behaviors within one-minute intervals (see fig. 10.2). The initials of the students' first names are used: Chuck (C), Dennis (D), Garin (G), and Razzie (R). Within each time interval, the order of each student's behavior is identified numerically. The asterisks indicate metacognitive behaviors.

By referring to the chart, it is possible to get a picture of the nature of the communication that took place within the group. This can be demonstrated by examining the chart during each of the first four minutes of the problem-solving endeavor in which the students were engaged.

During the first minute (i.e., 0–1), the students were presented with the graph (see fig. 10.1) and the task (i.e., to draw a picture of what the graph would look like if it were to continue from January through May). Razzie read the problem while Chuck, Dennis, and Garin listened (R1, C1, D1, and G1, respectively, in fig. 10.2). Garin proceeded to dominate the initial discussion as he attempted to understand, analyze, and explore how the problem might be solved (G2*, G3*, G4*). This seemed to give Razzie and Dennis ideas for participating with Garin in trying to understand and analyze the problem (R2*, D3*), while Chuck watched and listened (C2).

During the second minute (i.e., 1–2), the students took turns talking and listening to one another's ideas (C1, D1, G1, R1). Razzie started to analyze the average time of sunset (R2*), and each student was exploring possible patterns in time differences. Razzie began to construct a grid for the graph (R4). In his analysis, Dennis directed everyone's attention to a particular point on the graph, posing the question, "What happened here?" (D2*).

Episode or Stage

Episode or Stage	1	2	3	4	5	6	7
Read	C_1						
	D_1						
	G_1						
	R_1						
		C_3^*					
Understand	G_2^*	G_3^*					
	R_2^*		R_2^*	R_2^*		R_2^*	
Analyze			C_2^*	C_2^*			
	D_3^*		D_2^*	D_2^*			
	G_3^*		G_1^*				
		R_2^*					
Explore		C_2^*					
		D_2^*					
	G_4^*	$G_2^* \; G_4^*$					
		$R_3^* R_4^*$		R_3	R_3^*		
Plan				G_1^*		G_3^*	
Implement				C_3	$C_1^* \; C_3^*$	$C_1^* \; C_3^*$	C_2^*
				$D_3^* D_4$	$D_1 D_2^* D_3$	$D_1^* \; D_2$	$D_2^* D_3^*$
				C_3	G_1^*	$G_1^* G_2 G_4$	$G_1 \; G_2^*$
Verify							D_4^*
							$G_4 \; G_5^*$
Watch and Listen	C_2	C_1	C_1	C_1	C_2	C_2	$C_1 \; C_3$
	D_2	D_1	D_1	D_1	D_4	D_3	$D_1 D_3$
		G_1		$G_2 \; G_4$			G_3
		R_1	R_1	R_1	R_1	R_1	R_1

Fig. 10. 2. Sample coding: An analysis of six minutes of interaction amoung four students, C, D, G, and R, fifth graders at the Louis Armstrong Middle School, East Elmhurst

During the third time interval (i.e., 2–3), Garin did most of the talking while the others listened (C1, D1, R1). His insightful comments seemed to contribute to the students' discussion, which was focused on trying to understand and analyze the task (G1*, R2*, C2*, D2*).

During the fourth time interval (i.e., 3–4), Garin suggested a plan (G1*) to start in June at 8:30 and "keep on subtracting 20 you will probably get your answer for every month, maybe." Razzie was still trying to understand the task (R2*). Chuck and Dennis continued to analyze the situation (C2*, D2*), and Dennis noticed that there was no consistent pattern. Through their interaction, Dennis and Chuck realized that it might be worthwhile to try Garin's plan. Therefore, they attempted to implement Garin's plan (C3, D3*, D4, G3), even though it was based on the faulty idea that there was a regular pattern of time differences.

During the fifth and sixth time intervals (i.e., 4–6), it is clear that most of the students were engaged in the implementation of Garin's previous plan. The many asterisks indicate that the students were monitoring the work they were doing in trying to implement the plan. That is, they were listening to, responding to, and building on one another's ideas. It is interesting to examine the protocol from which this was coded. The excerpt from the transcript for time intervals 4-6 is given below.

Excerpt from transcript for time intervals 4–6

32. *Garin:* One hour and 20 minutes, one hour and 20 minutes, I'm right. So then it goes lower over here then one hour right? Yeah.

33. *Dennis:* So in between here is 5, 10, 35, and hour 20, and one hour.

34. *Garin:* So you want, so you want to do the same pattern over here it's over here like over, look. So let's do it here. (*Garin works on the paper in the center of the table.*)

35. *Chuck:* What time are we gonna start from?

36. *Garin:* Well are we gonna just...

37. *Dennis:* We have to figure out 5:20 to 30. (*Dennis points to paper in center of table.*)

38. ?: Let's see maybe starts it around,

39. *Chuck:* By the time we get to [inaudible] we're going to be down to like 3 o'clock. It gets dark at 3 o'clock? Be for real.

40. *Dennis:* No, because then you have to like start making the time go on higher. (*He makes small circular motions with hand and pencil.*)

41. *Garin:* (*hits Dennis's arm*) Oh, that's right, it goes look like this, listen, 8:30, subtract (*pause*) then it will probably go down. (*He works on paper in center of table.*)

42. *Dennis:* The difference in between the time gets larger and it also gets smaller, it starts decreasing. (*He points to paper in center of table and taps pencil.*)

43. *Garin:* All right, wait.

44. *Chuck:* Maybe it gets dark at 9 o'clock.

45. *Garin:* (*to himself*) 8:30.

46. *Dennis:* 9 o'clock?

47. *Chuck:* Yea, 9 o'clock. (*Garin works independently on paper in center of table.*)

48. *Razzie:* 9?

49. *Garin:* Look, look, wait, look, where is yours, look at this, would you listen? Look it goes down every 5, 10, 35 so why don't you just do this—subtract there from 5 and do it like, this is like your middle number, you got it? (*He taps desk with finger.*)

Note that in the interval 4–5, Chuck posed the question about the start time (C1*, see line (35)). The discussion about the start time continued to the point at which Chuck made the important observation that questioned the reasonableness of the sun setting at 3:00 in the afternoon (C3*, see line (39)). Chuck's revelation was a turning point in the solution process and led Garin to modify the plan (G3*, see line (40)) in interval 5–6. With the exception of Razzie, all the members of the group actively contributed to the successful completion of the task.

The group's solution is illustrated in figure 10.3. The students successfully completed the task, showing the continuation of the average-time-of-sunset graph from January through May.

As students worked in the small group to complete the graph task, the significance of the interplay of cognitive, metacognitive, and watch-and-listen behaviors was revealed, which supported the work of Artzt and Armour-Thomas (1992). Of the eighty behaviors coded during the seven-minute interval reported in figure 10.2, 47.5 percent were metacognitive, which indicates a substantial amount of control and regulation on the part of the group members in helping to guide and direct the task to a successful completion. This is a typical feature of expert problem-solving behavior (Schoenfeld 1987). Thirty-three percent of the behaviors coded were watch-and-listen behaviors. This indicates

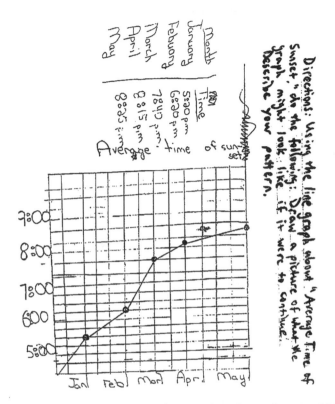

Fig. 10.3. The group's extension of the "average time of sunset" graph, which required reading beyond the data. The group was composed of students C, D, G, and R.

that a high degree of communication was necessary for effective small-group functioning. The analysis of the interaction revealed how students' metacognitive behaviors facilitated the building of ideas. Cognitive behaviors accounted for the smallest amount of coded behaviors (i.e., only 20 percent). This implies that the group worked efficiently, listening to, monitoring, and evaluating the different ideas presented for accomplishing the task without working mindlessly and going off on individual tangents.

The overall sequence of episodes illustrates the recursive nature of effective problem solving. The members of the group began by reading the directions and attempting to understand the task at hand. Next, some exploration and analysis led to a plan that was implemented. The plan was then revised as the students continued to implement the revision. Finally, the group went on to verify a solution (see fig. 10.2). This recursiveness is another characteristic of expert problem-solving behavior (Schoenfeld 1987).

FINAL REMARKS

The group highlighted in this discussion demonstrated the type of communication that is being recommended for improved learning in mathematics (NCTM 1991). Not all students, however, work so well when placed in a small group to solve a problem. The effectiveness of the small-group strategy is highly dependent on the nature of the assigned task (Cohen 1994). Note that the graphing task was of interest to the students in that it was related to their everyday experiences and it was challenging. These conditions are necessary for lucrative communication to take place within a group (Artzt and Newman 1997).

Another important contributor to the effective communication that took place within the group was the combined and individual characteristics of its members. That is, the group formation was based on the teacher's knowledge of the children's personalities and their intellectual capabilities. When they form groups, it is essential that teachers take into account the individual personalities and preferences of the group members and arrange for heterogeneity of ability (Artzt 1994). In this way, the chances will be maximized that students will partake in the give-and-take communicative behaviors that generate the kind of problem-solving behaviors that mirror those of an expert problem solver.

The results of this study confirm the idea that the small-group setting offers a fertile environment in which rich communication about mathematics may take place. That is, the nature of the communication that can occur when students try to solve a problem within a small group mirrors the monitoring and regulating strategies that experts reveal in their self-talk during solo problem solving. According to Vygotsky (1934/1986), the importance of having students participate in such a social setting is that the behaviors that occur within the group will later be transferred by the individual students when working by themselves. For the classroom teacher, this result implies that much can be gained by having students work in effective problem-solving groups.

REFERENCES

Artzt, A. F. "Integrating Writing and Cooperative Learning in the Mathematics Class." *Mathematics Teacher* 87 (1994): 80-85.

Artzt, A. F., and E. Armour-Thomas. "Development of a Cognitive-Metacognitive Framework for Protocol Analysis of Mathematical Problem Solving in Small Groups." *Cognition and Instruction* 9 (1992): 137–75.

Artzt, A. F., and C. M. Newman. *How to Use Cooperative Learning in the Mathematics Class.* 2d. ed. Reston, Va.: National Council of Teachers of Mathematics, 1997.

Cohen, E. G. "Restructuring the Classroom: Conditions for Productive Small Groups." *Review of Educational Research* 64 (1994): 1–35.

Curcio, F. R. "Comprehension of Mathematical Relationships Expressed in Graphs." *Journal for Research in Mathematics Education* 18 (1987): 382–93.

Kouba, V. L., C. A. Brown, T. P. Carpenter, M. M. Lindquist, E. A. Silver, and J. O. Swafford. "Results of the Fourth NAEP Assessment of Mathematics: Measurement, Geometry, Data Interpretation, Attitudes, and Other Topics." *Arithmetic Teacher* 35 (1988): 10–16.

National Council of Teachers of Mathematics (NCTM). *Curriculum and Evaluation Standards for School Mathematics.* Reston, Va.: National Council of Teachers of Mathematics, 1989.

———. *Professional Standards for Teaching Mathematics.* Reston, Va.: National Council of Teachers of Mathematics, 1991.

Schoenfeld, A. H. "What's All the Fuss about Metacognition?" In *Cognitive Science and Mathematics Education,* edited by A. H. Schoenfeld. Hillsdale, N.J.: Lawrence Erlbaum Associates, 1987.

Vygotsky, L. *Thought and Language.* Cambridge: MIT Press, 1934/1986.

11

Communication and Learning in Small-Group Discussions

Kaye Stacey and Anne Gooding
University of Melbourne

ANY teachers in Australia organize their classes into small groups of students working together. Teachers find that this arrangement meets a number of their goals. It enables all students to participate actively in a lesson, it reduces the number of individual requests by students for help from the teacher, and it enables students to help one another if students who might need help are teamed with students who can provide it. The purpose of the study reported here was to investigate the patterns of oral communication associated with successful learning in small groups. The learning task selected was an activity designed to correct misconceptions about division and the ways to write about it. The criterion for successful learning was improvement on a written test given before and after the activity. The learning task differs from other group-learning studies reported in the literature because its goal is not to teach new information or new skills but to reduce or eliminate misconceptions that children had developed. Of special interest was the possible discovery whether effective communication patterns in this situation would be the same as the patterns found to be effective in the acquisition of new learning.

This chapter contains material from Gooding and Stacey (1993a) and appears here with the permission of the publishers of the *Mathematics Education Research Journal*.

The authors thank Mollie McGregor for her advice on the manuscript. Since the preparation of this report, further studies of group interaction using a similar task have been undertaken. They are reported in Gooding, A. "The Learning Occurring in Cooperating Groups in Primary School Mathematics." Unpublished Ph.D. thesis, University of Melbourne, Australia, 1997.

There is a rich and diverse literature on the effectiveness of small groups in many curriculum areas, including mathematics. When mathematics achievement in the various cooperative schemes is compared to achievement from whole-class or individual learning, only a minority of the published studies have shown a statistically significant difference. Where significant differences have been found, however, they have usually favored the cooperative schemes (Davidson and Kroll 1991). This result indicates that the factors that make peer interaction a valuable device for learning are not automatically present when the classroom is organized around learning in groups.

Of particular relevance here are the studies that have examined the internal dynamics of cooperative groups and how they are associated with successful learning. In broad terms, it is clear from the literature that when interaction patterns of groups that learn effectively are compared with those of ineffective groups, students are seen to interact more with one another, interact more with the task, and use more cognitive strategies (Sharan and Shachar 1988; Webb 1991). Webb reviewed seventeen studies that examined verbal aspects of peer interaction. She reported that *elaboration of help* given was an essential feature of the peer interaction. Students who gave high-level elaboration (e.g., by explaining to their teammates and giving extended answers) showed higher achievement than those who did not. Webb used partial correlations in support of her claim that the high-level elaboration caused the high achievement rather than the alternative possibility that initially high achievement made the elaboration possible. Not receiving help when requested is negatively associated with learning. It would be expected that receiving explanations from other group members should be beneficial for learning. However, Vedder (1985) reveals that receiving explanations is sometimes effective for learning and sometimes not. It depends on such factors as whether the explanation is understood and whether the target student has, and then actually takes, an opportunity to use the explanation to solve the problem. The present study offers examples of all these instances.

For many students, learning in mathematics is plagued by persistent misconceptions, which are often untouched by traditional teaching methods. As yet, published research on the processes involved in group learning has studied only the acquisition of new knowledge. Sharan and Shachar (1988), for example, studied groups of students working together to learn material in history and geography for a class test. However, it seems likely that the reduction of misconceptions may be one of the most useful functions of group learning in mathematics because of the potential for the group situation to promote and then resolve cognitive conflict. One researcher investigating this possibility is Bell (1993), who has developed the diagnostic teaching methodology. This approach uses tasks designed specifically for groups to reduce misconceptions through small-group and class discussion. This chapter deals with a card-placing activity for small groups, which Bell found to reduce misconceptions effectively when supplemented by a class discussion. By using a pretest-posttest methodology, Bell showed that misconceptions were reduced in the class as a whole and that this effect persisted in the long term. However, the way in which the group processes contribute to the enhanced outcome has not been investigated for tasks of this nature. This chapter reports such a process-outcome study. The process (here, verbal interaction) occurring during each small group's discussion of the task is analyzed so that its impact on the learning that takes place (the outcome) can be examined. In this way, desirable process factors can be identified, and the means by which they are mediated by characteristics of both the students and the content can be studied. Good, Mulryan, and McCaslin (1992) suggest that much more research of this nature is needed.

Specific questions addressed in the study include the following: Do all children learn in groups? Is it possible to identify characteristics of effective and ineffective groups, as well as to distinguish between individuals who will improve and those who will not? Are some children limited by inadequate background knowledge or can the resources available in a heterogeneous group overcome this difficulty? What are the nature and content of the children's talk and how do they relate to what the children learn? How do children interact when they teach each other? How do they help each other learn mathematics? Many aspects of these questions are also raised by the work of Curcio and Artzt (chapter 10 in this volume), who studied the effects of small-group interaction on the comprehension of graphs by fifth graders.

The present study also documents, more directly than previous studies have done, how the effectiveness of the discussion relates to the subject matter of the children's talk. Striking differences have been found in the ways that children attend to the content.

METHODOLOGY

Choice of Task

Concepts and notation related to division were chosen as the topic of the discussion because the difficulties are widespread across the age range from ten years to adulthood. They are persistent, important to address, and well documented (Ball 1990; Hart 1981; Tirosh and Graeber 1990). Bell (1986) refined a task to help children learn about division of a small number by a larger number and sort out the confusions that arise from the different order of divisor and dividend associated with each of the signs used in elementary school for division, that is, ⌐ and ÷. These signs have no well-known names, so the names given by Cajori (1928), "Rahn's sign" for ÷ and "lunar sign" for ⌐ , have been used. In Australian schools, where this study was carried out, and elsewhere in the English-speaking world, children learn to read and write division with Rahn's sign from about age 7. For example, to divide 12 by 3 they write 12 ÷ 3. Later, when a formal setting is required for the division algorithm, at about age 10, children begin to use the lunar sign. For example, to divide 12 by 3 they write $3\overline{)12}$. Although the order of the operation for Rahn's sign and the lunar sign are opposite to each other (e.g., 16 ÷ 20 = 0.8, whereas $16\overline{)20}$ gives the answer 1.25), this does not generally cause any confusion at first because students are only required to divide a larger number by a smaller number. Given the numbers 16 and 20, they would assume that they should divide 20 by 16. Problems do not emerge until they need to divide a smaller number by a larger number and a choice of order must be made.

In the task, a group of children have to place thirty-six cards in appropriate places on a grid drawn on a rectangular board. Figure 11.1 shows the correctly completed placement of cards on the board.

There are three pairs of division questions. One question of each pair requires a larger number to be divided by a smaller number (e.g., $6\overline{)12}$ and 12 ÷ 6), and the other question requires the much harder reverse situation with the same numbers (e.g., $12\overline{)6}$ and 6 ÷ 12). In this chapter, the former divisions are called larger/smaller divisions, and the latter divisions are called smaller/larger divisions.

Procedure

Groups of four children in grades 5 and 6 were videotaped while working on the activity. The children were presented with the board, which had headings and a selection of the cards (those in boldface type in fig. 11.1) already in place. When they had placed the remaining cards to complete the board, they were asked to compare their board with one that was nearly correct and said to have been organized by children at another school. This gave them an opportunity to correct their own board and discuss any remaining misconceptions.

EXAMPLE	WORDS	÷	ANS	)‾	ANS
8 apples are shared between 2 boys. How many apples does each boy get?	**8 divided by 2**	$8 \div 2$	4	$2\overline{)8}$	4
2 apples are shared amongst 8 girls. How much apple does each girl get?	2 divided by 8	$2 \div 8$	1/4	$8\overline{)2}$	1/4
You have $12. Each present costs $6. How many presents can you buy?	12 divided by 6	$12 \div 6$	2	**$6\overline{)12}$**	2
What is 6 divided by 12?	6 divided by 12	$6 \div 12$	1/2	**$12\overline{)6}$**	1/2
4 kilometers split into 1/2-kilometer sections. How many sections are there?	4 divided by 1/2	**$4 \div 1/2$**	8	$1/2\overline{)4}$	8
1/2 a kilometer split into 4 sections. How long is each section?	1/2 divided by 4	**$1/2 \div 4$**	1/8	$4\overline{)1/2}$	1/8

Fig. 11.1. The completed board showing all thirty-six cards correctly placed (from Bell 1986).

The videotapes, including all the discourse and gestures that indicated the order of division operations, were transcribed. Two coding schemes, described subsequently, were used to analyze the discourse. A pretest was administered one week before the activity, and the posttest was administered three weeks after it. A relatively long-term learning effect was therefore being sought. Coded discourse was then related to the results of the tests.

Several schemes for coding the verbal interactions were tried. In the first instance, it was decided to concentrate on aspects of the interaction rather than on the content. Several schemes used in previous process-outcome studies were not regarded as suitable because they focused on linguistic form rather than on interaction. The most satisfactory coding scheme was developed from one used by Sharan and

Shachar (1988). These authors had withdrawn children from a classroom to work in groups of six to discuss a history or geography topic while they were being videotaped. Their coding was derived empirically from the data. They used, as the unit of analysis, "turn talking": an utterance of any length and grammatical form that is produced by one speaker without interruption. The Sharan and Shachar scheme described how students used language in interaction rather than what they said about the subject. Turns talking were coded as *focused interactions* (e.g., a request for clarification, agreement, or disagreement) and *cognitive strategies*, (e.g., explanation with evidence, unstructured idea or thinking out loud, concrete examples). The coding in this study was modified by combining and selecting the relevant categories of focused interactions and cognitive strategies according to the statements made during discussion of the mathematics task. The coding categories can be seen in figure 11.2, as well as some examples of the coding.

After the analysis of discourse according to the type of interaction was completed, a second scheme, developed precisely to capture the content difficulties of this task, was used to record the content of children's discourse in six categories. In this situation, short continuous episodes of speech about one decision was the unit of analysis. Examples are a short series of turns talking related to the placing of one card or a query about the placement of an earlier card with a subsequent discussion, involving several turns talking, about whether to move it. Episodes were categorized as to whether they involved smaller/larger divisions or larger/smaller divisions and whether these were correct or incorrect. Correct larger/smaller divisions were plentiful in all groups and yielded little information, so this category was not subsequently used. Discussions that indicated an awareness that the order of symbols in a division mattered in some—although not necessarily correct—way were coded as identifying noncommutativity. Episodes in which the order of division for either or both signs was discussed were also recorded, with correct and incorrect discussions separately.

The study by Curcio and Artzt (in this volume) presents both similarities and differences in methodology to this study. Both are studies of groups of children engaged in solving a problem for about fifteen minutes. In both studies, the researchers videotaped and transcribed children's discourse and analyzed the individuals' talk separately. Although both studies examine cognitive outcomes, in this study a pretest-posttest methodology is to identify learning some time after the task has been completed. Whereas the coding scheme presented here records aspects of interaction, Curcio and Artzt have used, as primary

Asking questions
- of a previous speaker
- from own thinking or working
- reading a word problem

How many 1/2's in 4?
Who's got 2? That goes there.
1/2 km split into 4 sections. How long is each section? We've got to work that out.

Responding
- to a request for clarification

S1: How many 1/2's in 4?
S2: There are 4 wholes, right?

- agreeing
- disagreeing
- repeating

Yeah
You're wrong
S3: No, 8 goes there.
S2: No, 8 goes there.

Directing

Put it down.

Explaining with evidence

No, 4. How many 1/2s is 8?
(4 ÷ 1/2; Pointing left to right and replacing 1/8 card with 8)

Thinking aloud when reading or placing the cards

12 divided by 6. All right.
12 divided by 6 is there.

Proposing ideas

It's 2 so they're not the same answer so we can't put them in.

Commenting (affective)

We've finished. Done it.

Refocusing discussion

Yeah. That's what you're meant to do. You're meant to set it out under this (pointing to headings).

Fig. 11.2. Examples of coding-scheme categories and examples of the coding for interactive aspects

categories, aspects of thinking (understanding and analyzing) These are embedded within the present categories. The emphasis here is to identify components of interaction by one scheme and to discuss aspects of the content by the other, rather than identify the specific thinking skills being used.

RESULTS

Pretest and Posttest Results

The task was effective in promoting learning, since half the class showed improvement on the difficult smaller/larger divisions on the posttest after three weeks. Each test comprised nine questions, of which the two smaller/larger divisions were the target questions. Almost all students were correct on the larger/smaller divisions on both occasions. However, on the pretest, only four of the twenty-eight children got either of the two target questions correct. On the posttest, fifteen were correct on one target item and ten were correct on the other. Individual students who improved their score on the target questions were designated improvers. The others were designated nonimprovers, except for the one child who had all questions right on both tests and who was not designated. The division questions and the results for the pretest and posttest are given in table 11.1.

Table 11.1
Results on the Target Questions of the Pretest and Posttest (N = 28)

	Smaller/larger		Larger/smaller	
Pretest questions	6)3̄	4 ÷ 16	5)15̄	12 ÷ 3
Number correct	4	1	25	25
(*N* = 28)				
Posttest questions	6)3̄	5 ÷ 25	3)15̄	12 ÷ 4
Number correct	15	10	22	26
(*N* = 28)				

Effective and Ineffective Groups

Five groups were designated effective because their members increased their score on the two target questions by an average of half a question or more. The other two groups (one of boys, one of girls) were designated ineffective. The learning outcomes for individuals were mixed. Each effective group had at least two improvers. Every group had at least one nonimprover. One ineffective group had completed the board correctly, and one, incorrectly. It is interesting that the only group in which no one made gains contained the only member of the class who could calculate smaller/larger items correctly from the beginning. He did not explain his results to the others in his group, and they placed their cards with the least discussion.

Interactive Aspects of Discussion

The two ineffective groups had students taking the fewest number of turns talking, 174 turns and 297 turns, respectively. In contrast, the number of turns talking in effective groups ranged from 362 to 820 turns. Among the effective groups, there is no direct correlation between the number of turns talking and the amount of improvement. Further details are given in Gooding and Stacey (1993a).

The results of coding the transcripts with the modified version of the Sharan and Shachar (1988) scheme are shown in table 11.2, where the mean percentage of turns talking in each category are given both for individuals and for groups. The children in the ineffective groups interacted less than did those in the effective groups. This is so even though the use of the percentages in table 11.2 adjusts for the very marked differences in the amount of talk among groups. The lower level of interaction for ineffective groups is reflected in the small overall amount of talk, the greater percentage of talk in the thinking-aloud category, and the generally lower percentages in the categories, such as responding, that indicated intellectual interaction. Similar observations hold when the comparison is between individuals (improvers and nonimprovers) rather than between groups, although a little less strongly.

Effective groups read the questions on the example cards out loud, whereas members of the ineffective groups did not do this at all (see table 11.2). Members of effective groups and improvers from all groups gave more explanations with evidence and repeated one another's statements more frequently. This is a way in which children agree, possibly reflecting on one another's answers at the same time (Noddings 1985). Because the percentage figures given in table 11.1 adjust for the differing amounts of talk in the groups, the absolute numbers of explanations and repetitions are very much greater for effective learners.

Mathematical Content of the Discussion

In addition to the foregoing interactive discourse analysis, the content characteristics of the discussion were analyzed. Turns talking were classified (where relevant) as correct or incorrect smaller/larger division statements, correct larger/smaller divisions, correct or incorrect discussion of the order of division, and statements generally identifying the noncommutativity of division. No category of incorrect larger/smaller divisions appears because very few occurred.

Almost no explicit mathematical discussion took place in the ineffective groups. Neither ineffective group specifically discussed the order of division operation for either division sign, whereas the effective groups discussed this explicitly eight or more times each. The ineffective groups each made only five calculations of any sort. Each effective

Table 11.2

Average Percents of Turns Talking for Each Type of Interaction for Groups and Individuals

	Effective Groups $N = 5$	Ineffective Groups $N = 2$	Improvers $N = 14$	Non-improvers $N = 13$
Asking questions				
• of a previous speaker	2.4	3.0	2.2	3.0
• from own thinking or working	7.1	6.6	7.7	5.7
• reading a word problem	1.0	0.0	1.2	0.4
Subtotal	10.5	9.6	11.1	9.1
Responding				
• to a request for clarification	6.1	5.5	6.2	5.6
• agreeing	9.4	5.5	9.2	8.2
• disagreeing	13.9	11.6	13.2	14.2
• repeating	4.5	0.5	4.7	2.6
Subtotal	33.9	23.1	33.3	30.6
Directing	5.8	2.7	4.9	6.3
Explaining with evidence	7.9	4.8	8.0	6.5
Thinking aloud when reading or placing the cards	36.0	55.6	36.2	43.6
Proposing ideas	1.4	0.7	1.7	0.6
Commenting (affective)	1.6	2.3	1.7	1.7
Refocusing discussion	2.7	1.4	3.1	1.4

groups made at least twenty-seven calculations. Table 11.3 gives the medians of the number of instances identified in the talk of each group. The number of instances of talk in each category for the two ineffective groups was, with only one exception, always less than the corresponding number of instances for each of the effective groups. The probability that the low results of the ineffective groups would have been obtained if there was no real difference between the groups was calculated by the Mann-Whitney test to be less than 5 percent ($n1 = 5$, $n2 = 2$, $U = 10$, $p < .05$). At the individual, rather than the group, level the results are similar. In that case it was appropriate to use the t-test on means rather than the Mann-Whitney test on medians. The results are given in table 11.3.

Table 11.3

Average Number of Statements in Specific Categories of Mathematical Talk (Standard deviations are in parentheses)

	Correct smaller/ larger	Incorrect smaller/ larger	Correct larger/ smaller	Identifying noncom- mutativity	Correct discussion of order	Incorrect discussion of order
Groups						
Medians for						
effective groups	7.0	17.0	8.0	10.0	10.0	6.0
Medians for						
ineffective groups	0.5	4.0	0.5	0.5	0.0	0.0
Significance of						
difference	*	*	*	*	*	n.s.
Individuals						
Means for						
improvers	3.4	4.4	3.1	3.5	3.9	1.1
Standard deviations	(3.18)	(4.37)	(3.87)	(2.77)	(4.30)	(1.12)
Means for						
nonimprovers	0.5	2.9	0.8	0.8	0.2	0.6
Standard deviations	(1.59)	(3.17)	(1.48)	(1.31)	(0.42)	(1.00)
Value of t (d.f. = 25)	2.84	1.00	1.90	3.07	3.08	1.16
Significance of						
difference	**	n.s.	n.s.	**	**	n.s.

* significant at the 5 percent level
** significant at the 1 percent level
n.s. not significant

DISCUSSION

Interactive Aspects

The broad patterns of interaction that have been associated with higher achievement in groups in previous studies have also been found in this new setting. Members of effective groups interacted more. Generally they helped one another more by responding and explaining to a greater extent during the task. Improvers engaged in the activity by explicitly working out the division problems in the task. They used more specific mathematical talk than nonimprovers for every category. Moreover, when adjusted for the absolute number of turns talking through the use of a percentage scale, effective groups showed more interaction in every category other than the noninteractive thinking aloud. The coding of mathematical content and the coding of interaction patterns have affirmed, in this new setting, Webb's (1991) proposition that giving help to others at a high level of elaboration is related to achievement. Not surprisingly, we found that those students who did not participate in the activity did not learn. However, we were surprised to find such large and consistent differences among students in the amount of participation and interaction.

A more detailed study of the patterns of interactions among the children is presented in Gooding and Stacey (1993b). By tracing in detail the content of the conversations and keeping track of the knowledge exhibited by each student through his or her statements, questions, and card placements, it was possible to discover who helped whom. Generally, the learning that was shown from pretest to retention test could be traced to explanations that the children in effective groups gave one another. However, for students to learn from the explanations, it was essential for them to receive appropriate explanation when required and to use the new ideas, for example, to help others or place other cards. Practicing the ideas by using them helped children learn. However, the group dynamics were found to be different for each effective group.

Background Knowledge and Achievement

It is likely that a threshold of background knowledge was required for a student to benefit from participation in the task. Although the numbers of students are too small to draw a statistically tested conclusion, it is perhaps significant that all the nonimprovers got both questions wrong on the pretest. The study by Curcio and Artzt (this volume) gives yet another example of a student (Student R) with low background knowledge and low interaction. He interacts at their lowest level of "watch and listen": little learning results.

However, no simple relationship exists between the background knowledge of group members (as measured by the pretest) and improvement. Group 7 contained the only student who had completely correct answers on the pre- and posttest, yet the others in his group did not improve. There was relatively little discussion and interaction; this group had the lowest numbers of turns talking and only about two-thirds as much talk as the next most taciturn group. In contrast, group 4 was effective, but the answers its members gave on the pretest showed not even a primitive "does not go" awareness of order of division. To a question such as $5 \div 25$ they would give the answer 5, completely ignoring any order problems, and they would not even comment that 25 into 5 "does not go." Further investigation, including a detailed mapping of the prior conceptual understanding of students, is required here to see how the students did improve.

Thinking Aloud

Although the results of this study support the general conclusions about interaction and learning found in other studies, the pattern of interaction was not the same. The most noticeable difference from the frequency of codings given in Sharan and Shachar's study (1988) was the very much higher percent of unstructured ideas coded as thinking aloud (3 percent compared to an average of 41 percent over all groups here). Rather than explain this difference by pointing to modifications made to the coding scheme, we propose that this result reflects differences in the nature and difficulty of the task. The children in this study spent a greater percentage of their time thinking aloud because they were working to understand difficult ideas. The difficulty of the task is indicated by the fact that all groups, especially the effective groups, frequently made mistakes. Thinking aloud is similar to Pimm's (1987) notion of stating a problem over and over. Pimm considers that this process can help establish a mental image, access to which is necessary for a solution. He sees it as a positive, clarifying aspect of problem solving. Our data associate it with ineffective learning, however, perhaps because children who were less able to cope with the task were unable to exchange mathematical ideas purposely and explain them to one another clearly. Students' capabilities and background knowledge as well as the nature of the task influenced their behavior and learning.

FINAL REMARKS

The task proved highly effective for reducing a persistent error for half the class in less than half an hour a group. This was most impressive because the posttest was three weeks later. Differences were found in the ways in which children in effective and ineffective groups engaged in mathematical discussion. Additionally, although the setting was different in important respects from the settings used in other studies of peer learning, effective discussion shared many of the features found in other studies. Both those groups that were effective and those individuals who improved interacted more than the others, and they engaged at a higher level of elaboration. Ross and Raphael (1990) have suggested that relatively unstructured tasks may be the most effective for achieving higher level cognitive objectives. This may be one feature that makes Bell's task appropriate for reducing misconceptions.

The analysis of the mathematical content of the discussion showed substantial differences in engagement with the central ideas of the task between effective and ineffective groups. In summary, effective groups—

- talked more, with more mathematical content (some of which was wrong);

- explicitly discussed the central idea (order of division);

- worked together by reading the questions on the cards out loud and repeating one another's statements;

- proposed ideas, gave explanations with evidence, and refocused discussion more often;

- responded to the questions of others more.

Improvers made many mistakes in their mathematical talk. But they were able to learn even though they made mistakes and heard others make mistakes. It can be hypothesized that Bell's task succeeded in creating cognitive conflict, which brought their misconceptions out into the open where the children in effective groups could grapple with them. This mechanism will be investigated in a future study. A detailed analysis of background knowledge and the conceptual development of group members, errors made during discussion, and the resulting learning could substantially illuminate the processes of learning and of reducing misconceptions through discussion.

Children's discussion with their peers in small groups is widely considered to be important for learning, yet the mechanisms that make it effective are not well understood. In this study, a detailed analysis of the ways in which effective and ineffective groups interacted in a task focusing on correcting misconceptions revealed substantial differences in patterns and substance of communication. The magnitude of these differences had not been apparent to the researchers while watching the groups. If teachers are made aware of the characteristics of effective group discussion, they will be able to recognize groups that are unlikely to learn and then take appropriate action.

REFERENCES

Ball, D. L. "Prospective Elementary and Secondary Teachers' Understanding of Division." *Journal for Research in Mathematics Education* 21 (1990): 132–44.

Bell, A. "Diagnostic Teaching: Two Developing Conflict-Discussion Lessons." *Mathematics Teaching* 116 (1986): 26–29.

———. "Some Experiments in Diagnostic Teaching." *Educational Studies in Mathematics 24* (1993): 115–37.

Cajori, F. *A History of Mathematical Notations.* Chicago: Open Court, 1928.

Davidson, N., and D. Kroll. "An Overview of Research on Cooperative Learning Related to Mathematics." *Journal for Research in Mathematics Education* 22 (1991): 362–65.

Good, T. L., C. Mulryan, and M. McCaslin. "Grouping for Instruction in Mathematics: A Call for Programmatic Research on Small-Group Processes." In *Handbook of Research on Mathematics Teaching and Learning*, edited by D. A. Grouws, pp. 165–96. New York: Macmillan, 1992.

Gooding, A., and K. Stacey. "Characteristics of Small Group Discussion Reducing Misconceptions." *Mathematics Education Research Journal* 5 (1993a): 60–73.

———. "How Children Help Each Other Learn in Groups." In *Communicating Mathematics: Perspectives from Classroom Practice and Current Research*, edited by M. Stephens, A. Waywood, and D. Clarke, pp. 41–50. Hawthorn, Victoria: Australian Council for Educational Research, 1993b.

Hart, K. M. *Children's Understanding of Mathematics 11–16.* London: John Murray, 1981.

Noddings, N. "Small Groups as a Setting for Research on Mathematical Problem Solving." In *Teaching and Learning Problem Solving: Multiple Research Perspectives*, edited by E. A. Silver, pp. 345–59. Hillsdale, N.J.: Lawrence Erlbaum Associates, 1985.

Pimm, D. *Speaking Mathematically: Communication in Mathematics Classrooms.* London: Routledge & Kegan Paul, 1987.

Ross, A. J., and D. Raphael. "Communication and Problem-Solving Achievement in Cooperative Learning Groups." *Journal of Curriculum Studies* 22 (1990): 149–64.

Sharan, S., and H. Shachar. *Language and Learning in the Cooperative Classroom.* New York: Springer-Verlag, 1988.

Tirosh, D., and A. O. Graeber. "Evoking Cognitive Conflict to Explore Preservice Teachers' Thinking about Division." *Journal for Research in Mathematics Education* 21 (1990): 98–108.

Vedder, P. *Cooperative Learning: A Study on Processes and Effects of Cooperation between Primary School Children.* Westerhaven Groningen, Netherlands: Rijkuniversiteit Groningen, 1985.

Webb, N. M. "Task-Related Verbal Interaction and Mathematics Learning in Small Groups." *Journal for Research in Mathematics Education* 22 (1991): 366–89.

12

Mathematical Communication through Small-Group Discussions

Marta Civil
University of Arizona

Marta Civil
University of Arizona

T HIS chapter draws on one experience in a mathematics course for preservice elementary school teachers, where much of the in-class work was done by students in small groups. At a very general level, a first question is, What happens when you let a group of students discuss a piece of mathematics? In trying to frame this question better, this chapter explores several factors that seem to affect small-group communication. It addresses such questions as the following: What can be learned about the students' understanding of, and beliefs about, mathematics as they talk mathematics in small groups? What is the role of language? How do the group dynamics influence the discussion? What is the teacher's role? What role do the tasks play? To address these questions, two protocols of small-group discussions are analyzed. The analysis illustrates the role that these different factors play as well as their interaction in the communication process.

SETTING THE CONTEXT

The work reported here is part of a larger study on the understanding of, and beliefs about, mathematics held by eight preservice elementary school teachers (Civil 1991). The study took place during an eight-week summer course that met for two hours every weekday. So that entrance could be gained into the mathematical world of the students, I carried out the research and the instruction and observed how the students talked about, wrote, and did mathematics. The instructional approach was thus instrumental to the research objective. The classroom was organized as a mathematics laboratory with tables for the students to work in groups and with a variety of manipulative materials. Lecturing was kept to a minimum. The usual approach was to present the students with a series of tasks on which they were to work. They sat in small groups, usually two groups of four. They were encouraged to advance their own ideas and to construct their understanding of mathematics. By moving away from the lecture format to an approach that focused on students' ideas, I created an environment for the teaching of mathematics that was quite different from their prior experience. Their beliefs about the teaching and learning of mathematics were being purposefully challenged.

The theoretical framework for this study rested primarily on constructivist approaches to learning and teaching mathematics (Cobb, Yackel, and Wood 1992; Confrey 1983, 1987; Lampert 1988; Noddings 1990). The main targets of interest were the students' ideas about mathematics and their reasons behind these ideas. Since peer dialogues were instrumental to this study, research studies addressing social interaction and the development of understanding in small groups were particularly relevant to the work reported here (Bartolini Bussi 1991; Hoyles 1985; Pirie 1991; Wood and Yackel 1990).

For the overall study, various sources of data were used: observations, informal conversations, written homework, essays and diaries, audiotaped interviews with one or two students, and audiotaped small-group discussions. In preparing the tasks for the study, I considered three different types of tasks that were likely to yield differences in the discussions:

- Tasks based on problem-solving situations; for example, the typical "finding the number of handshakes in a room with 100 people" problem
- Tasks based on "things they have always known"; for example, investigating the reasons behind the procedure to multiply fractions
- Tasks aimed at creating cognitive conflict; for example, a task presenting an alternative algorithm for one of the four arithmetic operations. This category also included situations in which conflict was the result of students' differing approaches to a problem.

The primary source of data for the work reported in this chapter comes from the small-group discussions. All the audiotapes were transcribed and the transcripts read and reread in search of episodes that offered a window into these students' understanding of, and beliefs about, mathematics. The next section outlines some of the difficulties in trying to make sense of mathematical communication. A closer look at two of the protocols then follows.

ANALYZING SMALL-GROUP DISCUSSIONS

It is often difficult to make inferences from a given episode, and several instances occurred in which the students' talk was found to be unclear. Some possible explanations for their lack of clarity might include—

- a poor understanding of the topic under discussion;

- a lack of experience communicating mathematics;

- a belief that a specific vocabulary and symbolism must be used in explanations (Balacheff 1986);

- a consequence of the group dynamics (for example, some students were not always comfortable sharing their thinking with some of their peers).

The analysis of students' talk had to engage with such issues as group dynamics and the use of language—oral, written, and nonverbal communication—in addition to the students' beliefs and understandings. Figure 12.1 reflects this view of the situation. This diagram stresses the interplay of the four elements that were constantly present in the analysis of the protocols of these small-group discussions. Pirie and Schwarzenberger's (1988) study on whether mathematical discussion among students contributed to mathematical understanding provided a valuable insight into the analysis of discussions. Of special interest to this study were the possible relationships that may exist between two of the components that the authors present as part of their proposed framework for analyzing discussion episodes. These two components were the kind of language that students use (whether ordinary or mathematical) and the type of statements that students make (whether, to use the authors' terminology, *reflective* or *operational*). The authors associated reflective statements with relational understanding and operational statements with instrumental understanding (Skemp 1976).

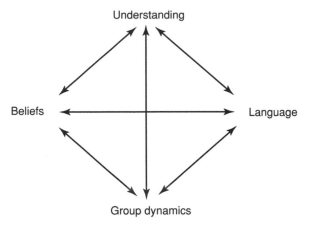

Fig. 12.1. Factors in a small-group discussion

A question such as "To what extent does lack of mathematical language inhibit reflective statements?" (Pirie and Schwarzenberger 1988, p. 469) reflects the kind of problem presented here because it stresses the interplay between language and understanding. This study is especially concerned with what sorts of inferences can be made—from the students' talk—about their understanding of mathematics. The next two sections focus on two protocols of small-group discussions that address the questions raised at the beginning of this chapter.

FORM OVER MEANING

In this task, the students were asked to find the following: In which base(s) (if any) is $(204)_? = (76)_{ten}$? They had been working on different bases for a few days. Although they were able to do such conversion tasks as $(354)_{eight} = (?)_{ten}$, they had not done much work with unspecified bases. That same day, during a discussion of one of the problems, they had been shown how to write $(143)_b$ as $1b^2 + 4b + 3$, but this method had not been stressed nor were they encouraged to use this line of thinking. In fact, I expected that they would approach this task by trying different bases, starting at $b = 5$.

Here is one excerpt from the dialogue among Ann, Betsy, and Lisa:

1. *Ann:*   This would be x and this would be 7 times x plus 6; do you think it will be $7x + 6 = 204$? So, you minus this....

2. *Betsy:*   Wait, I thought, you put the two-oh-four in letters.

3. *Ann:*   Ah.

4. *Betsy:* Because, don't you put it equal to what it is in base ten?

5. *Ann:* So, that should be x, you're saying.

6. *Betsy:* Uh, yeah.

7. *Ann:* Two-oh-four is the x; it would be two-oh-four x.

8. *Betsy:* But two-oh-four is really $2x^2 + 4$; I don't know whether this is right or not.

9. *Ann:* Why four?

10. *Betsy:* Oh, plus x plus four.

11. *Ann:* But, is it x or $2x$?

12. *Betsy:* It's 0, wouldn't it be 0 times x, which would make it 0?

13. *Lisa:* But 0 is not really 0, if ... uh....

14. *Betsy:* Oh, you're right, OK, OK.

15. *Ann:* So, you have $2x^2 + x + 4 = 7x + 6$?

16. *Lisa:* No, it equals 76, because it's in base ten already.

Ann was an older student, back in college after having raised a family. She seemed to be mesmerized by the idea of algebra. She felt very uncomfortable about her mathematics background, especially in algebra. She often turned to some of the other students whom she viewed as being above her because they knew algebra. Two of these students were Betsy and Lisa, who had more-typical mathematics backgrounds among elementary education majors. They were somewhat proficient at basic algebraic manipulations.

In this task, Ann took the initiative and led Lisa and Betsy to follow an algebraic procedure. At no point did they consider doing the problem any other way. Their approach was quite different from that of the other group, which opted to try different bases to find which one worked. In the excerpt presented, Ann starts by trying to follow what had just been shown to them on a previous problem, using the procedure like a template, focusing on surface features, and showing little understanding of what is going on. She immediately turns to follow Betsy's ideas about how to write the problem algebraically. Betsy herself is not too sure, and no argument is necessary to make her change her correct representation (8) to an incorrect one when Lisa says that "0 is not really 0" (13).

This statement, and the dialogue overall, illustrate these students' shaky understanding of place value. Although Betsy seems to have a clearer idea of how the algebraic procedure works, she hesitates all along (8, 10). One wonders how she would have done this problem had she been by herself. Would she not have actually succeeded in her algebraic approach? This may be one example in which the peer interaction did not contribute to a successful solution of the problem and did, in fact, lead students astray. However, the successful solution of a problem is not necessarily a sign of understanding. Betsy had described herself as good at memorizing and following rules. Hence, it is likely that by herself, she may have succeeded in patterning her work after the earlier problem and reached a correct solution through algebra. Had she done so, we would not have noticed her somewhat unstable understanding of place value.

This dialogue illustrates a focus on syntax at the expense of meaning. The students' main concern seems to be how to translate the problem into an algebraic form: "you put the two-oh-four in letters" ((2), operational statement). This dialogue supports the belief in algebra as a superior way to approach problems that was present throughout this study.

At the end of the excerpt presented, Ann, Betsy, and Lisa were confronted with solving $2x^2 + x + 4 = 76$, which they were not able to do. Meanwhile, Vicky, who was also in this group but had been working on her own, said that the answer was base six.

17. *Ann:* How did you do it?

18. *Vicky:* I went, seventy-six, you have four left over, seventy-two has to go into something twi.... It has to go into here to move it over, seventy-two divided by six would be twelve, which would give you the chips to move over.

19. *Ann:* So....

20. *Vicky:* Would give you twelve sets of six.

Vicky had visualized the problem in terms of chips and the trading mat that they had been using. She had first canceled out four chips in the units column, from 204, with four chips from 76, leaving her with $(200)_b = 72$. Then, she argued that b would have to be a divisor of 72 in order to move the chips to the b column (with none left over) and once more in order to move all the chips to the b^2 column: $72 \div 6$ gives twelve groups of chips, and dividing 12 by 6 again gives two groups. Thus, the base is six. Vicky seemed to have a clear understanding of the situation. Yet, her explanation (18) is likely to be hard to follow. As Pirie (chapter 1 in this volume) writes, we should not infer lack of understanding from unclear language. Vicky's language could not be described as either ordinary (natural) or mathematical. She was using the language of manipulatives. Verbalizing an action with chips—she was not using the chips, only visualizing them—is not an easy task. Many would find it easier to follow this argument: $2x^2 + 4 = 76$; hence, $2x^2 = 72$; hence, $x^2 = 36$, and thus $x = 6$. Mathematical language, especially symbolic language, does help in communicating mathematics. Vicky seemed to show an algebraic understanding in her talk (18), but this understanding was hampered by her lack of proficiency with the language of algebra and her firm belief that she could not do algebra, as the following excerpt shows.

21. *Ann:* What's the equation?

22. *Vicky:* I cannot do it with algebra.

23. *Betsy:* But here four was left over, but if you had like two two four, would you have done it the same way?

24. *Vicky:* I don't know.

25. *Betsy:* OK.

26. *Vicky:* You see, I use, most of my math is a process of elimination....

27. *Vicky:* If you can do it the algebra way and show me; I just can't. My problem is that I see the problem and I think back to doing it on the table.

Vicky did not seem satisfied with her method. Her prevailing thinking was that even though she could do the problems, she was not doing them the right way. It is unfortunate that Vicky's tone of helplessness and frustration in the last statement of this excerpt (27) cannot be conveyed. In this excerpt, Ann was still interested in how to do the problem algebraically, whereas Betsy challenged Vicky's method with another number. Was Betsy aware of the generalization power of an algebraic procedure versus what may have looked to her more like a trial-and-error method? Vicky did use her method to argue that there was no base b for which $(224)_b - (76)_{\text{ton}}$. But what did the students learn from this overall exchange? Once a solution was reached, Ann looked back at their algebraic work and made some progress toward a correct representation. That day, in her journal she wrote the following:

I love the formula $\_ x^2 + \_ x + \_ = \_$ tens (x is the base) to change one base to another. It is very clear to me to see it when I actually do a problem.

But, what can be said about their understanding? For example, did they walk away thinking that "0 is not really 0"? No teacher intervention was made during this discussion, and hence an opportunity to probe their understanding about place value may have been missed. This is one of the more difficult aspects of small-group discussions: What should be the role of the teacher? When should he or she intervene?

In this base example, many of the problems were not recognized until after the fact, and by then it was deemed too late to bring them up for further discussion. In the next task presented, the students were intentionally asked to look at one of their peers' approach to a problem because it reflected a typical difficulty in proportional reasoning.

EXPLAINING PROPORTIONAL REASONING

Earlier in the course, two students had agreed with a child's use of an additive strategy to solve a problem of proportional reasoning (Civil 1989). To encourage them to pursue this notion further, they were given the following task:

> If you need 1 1/3 cups of sugar and 4 cups of flour to bake a cake, how many cups of sugar will you need if you want to use 7 cups of flour?

Four of the five students who worked on this question came up with a correct solution to the problem. Carol used an additive strategy. She was asked to share her approach with the other students:

1. *Carol:* I wanted seven cups, so I put +3 and then I just added 3 here, so I got 4 1/3, but it's wrong, so I don't know why I'm explaining it to you.

2. *Betsy:* You want three more cups of flour, but you don't know that you want three more cups of sugar; that's the question—how many cups of sugar do you need.

3. *Lisa:* Yes, that's the unknown.

4. *Betsy:* See what I'm saying?

5. *Carol:* No.

6. *Betsy:* Because you are saying, I want three cups of flour and I want three cups of sugar....

7. *Carol:* But if you are tripling the recipe....

8. *Joyce:* You're not tripling.

9. *Vicky:* You're not multiplying, you're adding.

> (*Joyce mumbles something along the lines that if you were tripling, you would be getting twelve cups of flour.*)

10. *Lisa:* You are adding 3 here to equal 7; this is your unknown over here [ ]:

$$1\frac{1}{3} \qquad 4$$

$$[\ ] \qquad \frac{3+}{7}$$

You're saying, if you're getting seven cups here, this plus, you know whatever, it's going to equal, you don't know; you don't know it's going to end up to be 7; this isn't necessarily going to be 7, you may need....

11. *Betsy:* You say you are tripling, but you're not.

The dialogue for this task illustrates one of the concerns in small-group discussions with respect to group dynamics. For Carol, this approach to learning mathematics became increasingly painful as the following excerpt from her journal shows:

> Again, I am concerned about this nontraditional method of teaching. It seems to be a good thing for group cooperation and for higher ability math solvers, but not for people like myself.... What happens in the group dynamics is that those who understand, have background knowledge, etc. get better and people like myself get worse. I think it's a lot to ask of myself (at 24) and kids, very young ones, socially to appear as a constant failure to his/herself, peers, the teacher.

The goal in having Carol share her approach was to learn about the other students' understanding of proportional reasoning. Although they had successfully solved the problem, their understanding of the problem was not obvious. In the first phase ((1) through (11)), the four students took over and basically told her that she could not add 3.

Lisa's statement (10) deserves special attention, since she stayed with it for the rest of the discussion. She seemed to be interpreting Carol's strategy as doing

$$1\frac{1}{3} + ? = 7,$$

that is, as if Carol were saying that the target was to get seven cups of sugar ("you don't know it's going to end up to be 7; this isn't necessarily going to be 7").

Carol referred to "adding 3" as "tripling"—an interesting use of language but one that was not pursued at this time. Her peers discarded this by telling her that she was not tripling but adding. Carol's concept of ratio was rather unstable, as the following quotation shows. In response to "give two numbers which are in a 2 to 5 ratio," she said the following:

2 to 5? OK, a 2 to 5 ratio would be 4 to 10; wait a minute—plus 3; no, it would be 5 to 8.

Here she moved from a multiplicative procedure to an additive one. It is possible that her use of "tripling" was also subject to this instability, meaning sometimes adding 3, sometimes multiplying by 3. Carol was very concerned about what she perceived as her lack of mathematical vocabulary, as shown in this excerpt from her journal:

> A majority of my frustration arises from the fact that I do not have a "math vocabulary" and have a hard, almost impossible, time saying what I mean.

In the second phase of this dialogue, Carol was completely taken aback by the fact that Lisa was going from $n{:}m$ and $a{:}b$ to

$$\frac{n}{a} = \frac{m}{b}.$$

One can feel the force of this revelation for Carol, judging by her insistence on the fact that "I didn't know that a fraction was the same thing as a ratio."

12. *Carol:* So, how do you do it?

13. *Lisa:* OK, you are right in setting up ratios. OK, you have 1 1/3 cups of sugar to every 4 cups of flour and that's right (*writes*

$$\left. \frac{1\frac{1}{3}}{4} \right).$$

Now, you can set that equal to ?/7, like I said over here. You want to know how many cups of sugar it's going to take for every 7 cups of flour.

14. *Carol:* I had no idea that a fraction was like a ratio.

15. *Lisa:* You did have an idea, though; you just got confused; you had the right idea; these two dots mean ratio technically.

16. *Carol:* But I didn't know that a fraction was the same thing as a ratio.

17. *Lisa:* Well, you can set it up this way [with the two dots] if you want, if that's easier for you to look at.

18. *Carol:* But weren't you writing this as a fraction?

19. *Betsy:* To solve it.

20. *Lisa:* Yeah, to solve it, just to solve it.

21. *Carol:* I didn't even know that 3:4 is 3/4. I didn't know that until you just explained this.

In this excerpt, the goal was to show Carol a procedure for setting up this problem. Lisa was being very comforting all along (15, 17), trying to convince Carol that she was on the right track. Also, it should be noticed once again, a mastery of basic symbolic manipulation, such as that exhibited by Lisa and Betsy in setting the problem, helped the communication—provided that they were all familiar with this symbolic language. Carol, however, was not familiar with it. In fact, her comment "I had no idea that a fraction was like a ratio" opened up a whole new dimension. This reaction was not expected. What was Carol's understanding of ratios and fractions? An individual interview was set up to explore further her understanding of these concepts.

The third phase of this dialogue (not included herein) was an attempt to bring the students back to Carol's method and to a discussion of why adding 3 was not a correct way to solve the problem. Lisa insisted on her previous interpretation. Betsy insisted that "you are not adding three whole cups." Vicky's statement offered a new, potentially very powerful, look at the problem but remained ignored. Betsy and Lisa dominated the discussion.

Vicky: You're not dealing with whole numbers; you're saying that a part of a number equals a whole number, so you can't add three whole numbers to the partial numbers.

What did she mean? Once again, as was often the situation with Vicky, her verbal expression was not the clearest to follow, yet she was usually right on target. To understand her statement, we need to look at how she had solved the problem. Vicky had solved it in a visual way, an approach that was very different from that of her peers who had all set it up as an equality of ratios and followed an algebraic procedure. Vicky, instead, had drawn the cups of sugar and of flour and saw immediately that there was one cup of flour for every 1/3 cup of sugar; she argued then, that if one were to use 7 cups of flour, one would need 2 1/3 cups of sugar. This is what she meant when she said that "a part of a number equals a whole number": "a part of a number" refers to 1/3 and "a whole number" refers to 1. Through her drawing it out, she had brought the ratio to 1/3 to 1.

By the fourth phase it was felt that having the students explain why adding was not the correct reasoning was not going anywhere. Therefore, they were asked to talk about how they would explain how to do this problem to someone else. Lisa's strategy of changing the problem to whole numbers was an interesting approach that may have helped the others visualize the situation. Lisa's saying that "2 cups of sugar for every 4 cups of flour" (29) combined with Vicky's visual approach might have helped develop an understanding of multiplicative versus additive strategies. However, these thoughts can only be surmised, since Betsy and Lisa controlled the protocol. Lisa insisted on her interpretation of Carol's reasoning. She still thought that Carol suggested adding 3 because she wanted to get seven cups of sugar (38). Betsy's explanation was based on the idea that the ratios should be conserved (39, 41, 43), an idea that, it seems, was not at all clear in Carol's mind. In fact, it would appear that this is where Carol's understanding of ratios failed.

29. *Lisa:*   Well, OK, you could explain it to a kid or whatever, by, just cross out the 1 1/3 and make it easier—put a whole number there, like 2; OK, 2 cups of sugar for every 4 cups of flour; OK, try that—put that down on paper.

30. *Carol:* What am I doing?

31. *Lisa:*   Change the 1 1/3 to 2 cups instead.

32. *Carol:* So, I just make it 2 to what?

33. *Lisa:*   4....

34. *Carol:* 4, OK.

35. *Lisa:*   And you are going to end up with 7 cups of flour, OK. So, you are saying that if you add 3, you're going to end up with 7 cups of flour.

36. *Betsy:* So, add 3 to the other side....

37. *Carol:* Which is 5.

38. *Lisa:*   OK, so you don't get 7 cups.

39. *Betsy:* 2 is half of 4, right?

40. *Carol:* Uh, uh.

41. *Betsy:* Is 5 half of 7?

42. *Carol:* 5 would be, no....

43. *Betsy:* Right, so you know that's wrong.

44. *Carol:* Oh, OK.

45. *Lisa:*   Do you understand that?

46. *Carol:* Yeah, I understand that (*in a rather submissive tone*).

47. *Lisa:*   OK, it's just a little bit more confusing when you have a fraction in there, because it's harder to figure it out right in your head.

However, it is not certain that Carol really did understand. For her, it was once more an authority telling her that what she had done was wrong. She usually tried very hard to understand and was not willing to just accept. But by that point in this problem, she appeared to be ready to give up. The discussion approach that characterized this mathematics course was not welcomed by Carol, who quite often felt left out, as shown in a journal entry:

> I try to understand, and in class I listen and ask questions but most of the time I have absolutely no idea what is going on. And what my peers say to me sounds like a dialect of the Alaskan Eskimo.

In all fairness to the students, it should be noted that what they were being asked to explain is a difficult concept to which many researchers have devoted considerable effort. Of particular relevance to the topic of this chapter is the fact that the task appeared to belong more to the researcher than to the students. Of course, the same could be said about most tasks used in this study. But in many of the other tasks (e.g., the bases problem discussed earlier), the students had some control. They could make it into their own task (e.g., how to solve the base problem using algebra). But in the cups-of-sugar problem, they were pushed into addressing a conflict situation. It is not at all clear that for these students, Carol's use of an additive strategy was something that they were interested in discussing. It was thought to be noteworthy (by the researcher) because this pattern of reasoning is evident in many teaching situations in mathematics. The students, however, did not have this awareness. Many factors were at play in this task, a main one being the players themselves. Students were aware of Carol's plight in this course and may have been uncomfortable by having to talk to her about her mistake.

Final Remarks

In trying to understand mathematical communication through small-group discussions, I initially considered four factors: the students' understanding of the mathematics being discussed, the students' beliefs about mathematics, the effect of the group dynamics, and the role of language. As the study developed, two more factors—the teacher's role and the type of task being discussed—became necessary to the analysis. These six factors have been addressed in this chapter. The following paragraphs contain brief summarizing comments on each factor and raise some points that need further research.

The two protocols presented illustrate aspects of the students' understanding of place value, algebraic manipulation, and proportional reasoning. The discussion pointed out how some of the students' belief in algebra as a superior way to approach a problem may, in fact, have prevented them from looking at other approaches that could have shed light on the problem. Students seemed to be aware of the potential power of algebra and of the elegance and clarity (once mastered) of algebraic expression. Symbolic language often facilitates mathematical communication more than ordinary (natural) language does. For example, in the dialogues discussed, Vicky relied on a combination of natural and visual language to explain her thinking. Her explanations were almost always of this nature. Without taking the time to put them in context and reconstruct what she was saying, one would often miss her mathematical insight. And since Vicky was not particularly pushy and felt somewhat uncomfortable about her mathematics ability, her contributions to the group discussions sometimes went unnoticed. Although this situation improved as the course went along and students became more used to listening to one another and to working together toward a solution or an explanation, this is not to say that the issue of unconventional language had been settled. As Austin and Howson (1979) (citing Arnold [1973]) point out, "many teachers (and students) are unable to cope with formulations of ideas which are not expressed in the subject register and often fail to recognise their validity" (pp. 175–76). Many teachers feel uncomfortable when confronted by students' careless, or even incorrect, mathematical talk, especially when communication among themselves seems to be taking place successfully. A related question that needs further research is, therefore, What is the relationship between the ownership of an idea and the use of language?

In looking at the communication of mathematics in small groups, we must take into account the dynamics among the different members. In this study, the most mathematically productive discussions came from a group in which two of the members were used to challenging each other's point of view because they had been friends for a long time. Also, most students felt very comfortable working with two of the older students because they were very gentle and caring in their interactions. Although teacher interventions were kept to a minimum, the information gathered from these latter discussions was used to design new tasks to explore the students' understanding further.

A final element in small-group discussion is the role of tasks. The larger study (Civil 1991) seems to indicate that the problem-solving tasks were more conducive to a social construction of meanings, whereas the tasks on "things they have always known" were more illustrative of the students' beliefs, particularly about teaching. However, more research is needed to look at the effect on communication of different types of tasks. In this study the researcher had considerable control over the tasks. The tasks were hers, not the students'. Yet at times, the students transformed the given tasks and made them more their own. How this process affects communication in small groups is a question worth pursuing. Finally, it must be pointed out that these six factors are hard to consider in isolation. The protocols have illustrated some of the interactions among these factors. A systematic analysis of how these factors interact is a field for further research.

REFERENCES

Austin, J. L., and A. G. Howson. "Language and Mathematical Education." *Educational Studies in Mathematics* 10 (1979): 161–97.

Balacheff, N. "Cognitive versus Situational Analysis of Problem-Solving Behaviors." *For the Learning of Mathematics* 6 (1986): 10–12.

Bartolini Bussi, M. "Social Interaction and Mathematical Knowledge." In *Proceedings of the Fifteenth Conference of the International Group for the Psychology of Mathematics Education*, vol. 1, edited by F. Furinghetti, pp. 1–16. Assisi, Italy, 1991.

Civil, M. "Prospective Teachers' Conceptions about the Teaching and Learning of Mathematics in the Context of Working with Ratios." In *Proceedings of the Eleventh Annual Meeting of the North American Chapter of the International Group for the Psychology of Mathematics Education*, vol. 1, pp. 289–95. New Brunswick, N.J., 1989.

———. "Doing and Talking about Mathematics: A Study of Preservice Elementary Teachers." Ph.D. diss., University of Illinois, 1990. Abstract in *Dissertation Abstracts International* 51 (1991): 4050A.

Cobb, P., E. Yackel, and T. Wood. "A Constructivist Alternative to the Representational View of Mind in Mathematics Education." *Journal for Research in Mathematics Education* 23 (1992): 2–33.

Confrey, J. "Young Women, Constructivism, and the Learning of Mathematics." In *Proceedings of the Fifth Annual Meeting of the North American Chapter of the International Group for the Psychology of Mathematics Education*, vol. 2, pp. 232–38. Montreal, 1983.

———. "The Constructivist." In *Proceedings of the Eleventh International Conference for the Psychology of Mathematics Education*, vol. 3, edited by J. C. Bergeron, N. Herscovics, and C. Kieran, pp. 307–17. Montreal, 1987.

Hoyles, C. "What Is the Point of Group Discussion in Mathematics?" *Educational Studies in Mathematics* 16 (1985): 205–14.

Lampert, M. "The Teacher's Role in Reinventing the Meaning of Mathematical Knowing in the Classroom." In *Proceedings of the Tenth Annual Conference of the North American Chapter of the International Group for the Psychology of Mathematics Education*, pp. 433–80. De Kalb, Ill., 1988.

Noddings, N. "Constructivism in Mathematics Education." In *Constructivist Views on the Teaching and Learning of Mathematics, Journal for Research in Mathematics Education* Monograph No. 4, edited by R. B. Davis, C. A. Maher, and N. Noddings, pp. 7–18. Reston, Va.: National Council of Teachers of Mathematics, 1990.

Pirie, S. "Peer Discussion in the Context of Mathematical Problem Solving." In *Language in Mathematical Education: Research and Practice*, edited by K. Durkin and B. Shire, pp. 143–61. Milton Keynes: Open University Press, 1991.

Pirie, S. E. B., and R. L. E. Schwarzenberger. "Mathematical Discussion and Mathematical Understanding." *Educational Studies in Mathematics* 19 (1988): 459–70.

Skemp, R. R. "Relational Understanding and Instrumental Understanding." *Mathematics Teaching* 77 (1976): 20–26.

Wood, T., and E. Yackel. "The Development of Collaborative Dialogue within Small-Group Interactions." In *Transforming Children's Mathematics Education: International Perspectives*, edited by L. P. Steffe and T. Wood, pp. 244–52. Hillsdale, N.J.: Lawrence Erlbaum Associates, 1990.

13

Formats of
Argumentation in the
Mathematics Classroom

Götz Krummheuer
Freie University, Berlin

T HE theoretical description of the teaching-learning process in
mathematics education is dominated by concepts of psychology.
Here, an alternative (micro) sociological approach proposes that
children's participation in a social process of establishing formats of
argumentation also contributes to the learning of mathematics.
Several episodes of elementary school mathematics classes illustrate
this statement.

In this chapter, learning is understood as an internal cognitive
process situated in a social context, in which the learner participates in
processes of social interaction.

> What a learning child is doing is participating in a kind of cultural geog-
> raphy that sustains and shapes what he or she is doing, and without
> which there would, as it were, be no learning. (Bruner 1990, p. 106)

One central aspect of this participation of a learning child in the "culture of the mathematics classroom" is its contribution to processes of
"collective argumentation." The so-called teaching-learning process is
to be understood, therefore, as a social event in which certain "formats" of collective argumentation are initiated, generally by the
teacher, and stabilized by negotiation among all participants in the
classroom setting.

In the first section, the concept of "collective argumentation" is
explained in more detail. Second, the concept of the "format of collective

223

argumentation" is clarified; finally, the theoretical framework is outlined.

THE CONCEPT OF COLLECTIVE ARGUMENTATION

A typical feature of teaching-learning processes is that the student's individual learning is embedded in a social process of explaining, clarifying, explicating, and illustrating. These interaction processes of arguing contribute to, initiate, orient, and evaluate individual learning. As Miller (1986, p. 248) says, "most individuals not only learn to argue, they also learn by arguing" (translated by the author of this chapter. See also Miller [1987]. For the reception of Miller's approach in the United States, see, for example, Rogoff [1990, pp. 178–80] or Forman [1991]).

If one uses the concept of argumentation in the field of mathematics, one might tend to bind it closely to that of proof. The analysis of argumentation in a classroom could then be misleadingly understood as a treatise on proof. Therefore, one should notice that the concepts both of argument and of argumentation need not be exclusively connected with formal logic as we know it from such proofs, or seen only as the subject matter of logic. Many human activities and efforts are argumentative but not logical in a strict sense. As Toulmin (1969) points out, if these formal logical derivations of conclusions were the only legitimate form of argumentation at all, then the domain of rational communication would be extremely restricted, and argumentation as a possible way of a communication, based on rationality, would be rather irrelevant (p. 40 f., p. 123 ff.).

Accordingly, Toulmin distinguishes between the concepts of "analytic" and "substantial" argumentation. A logically correct deduction, for example, contains in its conclusion nothing that is not already a potential part of the premises. It *explicates* certain aspects of the meaning of the premises by means of deduction. Such kinds of argumentation are analytic. In contrast, substantial arguments *expand* the meaning of such propositions insofar as they relate a specific case to them by actualization, modification, application, or all three. Thus, substantial argumentation is informative in the sense that the meaning of the premises increases or changes by the application of a new case to it, whereas analytic argumentation is tautological; that is, a latent aspect of the premises is visibly elaborated[1]. Usually this kind of substantial

---

1. See also Wittgenstein (1963): "The propositions of logic are tautologies. Therefore the propositions of logic say nothing (They are the analytic propositions.)" (§6.1, 6.11)

argumentation does not have the logical stringency of a formal deduction, which, however, should not be understood as a weakness but rather as evidence for the existence of problems that are not accessible to formal logic.

The only arguments that can be judged fairly by deductive standards are those "held out as and intended to be analytic, necessary and formally valid" (Toulmin 1969, p. 154).

As Toulmin strongly emphasizes, substantial argumentation should not be subordinated or related to analytic argumentation, in the sense that the latter forms the ideal type of arguing, whereas there is always a "logical gulf" in a substantial argument (cf. p. 234). Substantial argumentation has a right of existence of its own. By means of substantial argumentation, a statement or decision is *gradually* supported. This support is provided neither by a formal, logical, necessary conclusion nor by an arbitrary edict, like declared self-evidence, but is achieved by the accomplishment of a convincing presentation of backgrounds, relations, explanations, justification, and qualifiers.

> The very nature of deliberation and argumentation is opposed to necessity and self-evidence, as no one gets into deliberations if the solution is necessary or argues against what is self-evident. The domain of argumentation is the field of the credible, the plausible, the probable, to the degree that the latter eludes the certainty of calculation. (Perelman and Olbrechts-Tyteca 1969, p. 1)

This distinction helps clarify the conceptual framework chosen here for the analysis of argumentation in mathematics classroom situations in primary school education. It is the substantial argumentation that is seen as more appropriate for this purpose. Again this does not imply that such argumentation has to be judged as poor or weak. At this point at least two reasons are given to support this decision.

First, children generally do not act on the level of an axiomatic mathematical system. Primary school children's mathematical knowledge is more at an empirico-theoretical level (Struve 1990), and their mathematical statements are based on real experience (Cobb 1994), for example, counting on fingers, concluding from an actually existing concrete embodiment, or taking evaluations of an authoritative person for granted.

Second, at the primary level, children usually do not exclusively draw analytic conclusions. Studies about the ontogenesis of the ability of arguing put forward the claim that deductive conclusions—for example, the part-whole inclusions—appear relatively late, developmentally speaking, in their fully developed verbalized form. In particular, Piaget's

theory of the development of causal thinking claims that deductive conclusions are based on a process of several prior stages and, even at the
latest stage, that of "representative causality," some residue of the previous stages remains (Inhelder and Piaget 1985; Keil 1979, 1983; Piaget
1928, 1930, 1954; Siegal 1991; Völzing 1981). (A summary of Piaget's
approach can be found in Siegal, pp. 39–45.)

Empirically, the concept of argumentation will be related to those
interactions in the observed classroom that have to do with an intentional explication of the reasoning leading to a solution during or after
its elaboration. Argumentation is generally understood here as a specific feature of social interaction.

If one or several participants assert "4 times 10 equals 10 times 4" or
"31 + 19 is the same as 32 + 18," they not only produce a sentence;
they also make a declaration inasmuch as they *claim* such a statement to be valid. By proposing it they are indicating not only that they
are trying to act *rationally* but also that they could *establish* this
claim in more detail, if desired. Usually these techniques or methods
of establishing a statement are called *argumentation*. Thus a successful argumentation refurbishes such a challenged claim into one
that is based on a consensus or is acceptable for all participants
(Kopperschmidt 1989, p. 24 ff.).

Within a classroom, argumentation generally takes the form of a
direct face-to-face interaction rather than that of a monologue. Because
of the emergent nature of social interaction, argumentation is usually
accomplished by several participants. In this situation, we speak of
"collective argumentation" (Krummheuer 1992; Miller 1986, 1987).
Moreover, the development of a (collective) argumentation need not
proceed in a harmonious way. Disputes may arise in various episodes
of argumentation, which leads to corrections, modifications, retrac-
tions, and replacements. Thus the set, or sequence of statements, of
the final agreed-on argumentation is shaped step-by-step by overcoming controversies. The result of this process can be reconstructed and
is called an "argument." To reconstruct an argument sufficiently, the
participants or the observer must keep in mind these interaction
processes.

A short episode illustrates the process of a jointly produced argumentation. Two second graders, Andy and John, work together in a peer
group and are to solve the following set of problems[2]:

---

2. For more details see Krummheuer and Yackel (1989, p. 116, vol. 3); for more information about the underlying Purdue Problem-Centered Mathematics Curriculum
   Project, see, for example, Cobb, Yackel, and Wood (1989).

$$2 \times 4 = \_ \quad 4 \times 4 = \_ \quad 5 \times 4 = \_ \quad 10 \times 4 = \_ \quad 9 \times 4 = \_$$
$$8 \times 4 = \_ \quad 8 \times 5 = \_ \quad 7 \times 5 = \_ \quad \_ \times 5 = 30$$

These problems had been given on the day when the concept of multiplication had been introduced in terms of sets: p sets of q elements each have "$p \times q$" elements. The episode is that of the students' joint solution of the fourth problem, namely, $10 \times 4 = \_$.

1. *John:* Ooh! Just one more than that [5×4].

2. *Andy:* No. No way!

3. *John:* No, look. It's five more sets [of 4]. Look.

4. *Andy:* Yeah.

5. *John:* Five more sets than 20.

6. *Andy:* Oh! 20 plus 20 is 40. So it's got to be 40.

Here one can reconstruct how John's statement in line (1) is rejected by Andy (2). John's reaction in line (3) indicates that he now also disagrees with his first statement (of line (1)). Thus conjointly this first utterance is rejected. In line (3) John is already proposing a new statement with which Andy agrees in line (4). John then infers on this commonly accepted basis that 10×4 must be "five more sets than 20" (5). Andy seems to agree by explicating a further deduction from John's statement of line (3): "So 20 plus 20 is 40" (6). Finally he concludes that the result has to be 40 (6).

Both children are actively engaged in the creation of an argument that not only leads to a result but also explicates reasons for its correctness. One can also see that during the process of this collective argumentation, one can point to statements that do not find agreement among the interlocutors.

THE CONCEPT OF
FORMAT OF ARGUMENTATION

From the interactionist point of view (Turner 1988; Bauersfeld, Krummheuer, and Voigt 1985), a central question is how a process of collective argumentation is structured when a jointly accepted argument is successfully negotiated.

In the following section, this question is treated first by introducing the concept of "format" and second by reporting two episodes as an illustration.

This concept is introduced by Bruner in his later studies on language acquisition in children. His approach is based on the fundamental insight that the learning of language is inextricably interwoven with the learning of the situations of its usage. For example, the child not only learns the words necessary to ask for a cookie; she primarily learns the social situation of request, where these words fit in (see also Bruner [1983]).

> [I]n order for the young child to be clued into the language, he must first enter into social relationships of a kind that function in the manner consonant with the uses of language in discourse. (Bruner 1985, p. 39)

For Bruner this relationship is bound not only to the acquisition of one's native language. Mathematics classroom situations also need to be analyzed with regard to such social relationships that function in a manner compatible with the uses of mathematics among adults. In other words, participation in the processes of collective argumentation contributes to the learning of mathematics in classroom situations. Interaction formats are defined by Bruner in the following way (see also Bruner [1983, p. 8]):

> A format is a standardized, initially microcosmic interaction pattern between an adult and an infant that contains demarcated roles that become eventually reversible. (Bruner 1983, pp. 120–21)

A format regulates the steps of interaction. Specifically with regard to mathematics classroom interaction, it sets the sequence of statements of a collective argumentation in an order that has been officially declared appropriate. It does not determine the participants' definitions of the situation. It helps rather to *orient* the students' processes of constructing new meanings as well as to increase the chances of the creation of new meanings, which gradually *converge* to a pedagogically intended mathematical definition of the situation. (For the function of the orientation and convergence of formats of argumentation, see Krummheuer [1992]. Detailed discussions of the concept of pattern of interaction can be found in Voigt [1984, 1989, 1995].)

Formatted argumentations are often generated by participants of qualitatively different mathematical abilities—for example, when the teacher is involved. In this situation, the teacher usually proposes a guideline for the production of appropriate arguments. But this asymmetric pattern is not necessary for the accomplishment of a formatted argumentation. Also when students work together on several problems, which they solve in a similar way, a format of argumentation can emerge with all the described impact. It is a matter of empirical reconstruction to clarify the importance of symmetry or asymmetry in conjunction with the concept of format. Bruner's empirical data are gathered in the mother-child interaction as a basis for language acquisition in early childhood. Here the asymmetry is obvious. With regard to the numerous occasions of small-group work in mathematics classes, the issue of symmetry becomes more noteworthy.

Two episodes from the same German first-grade mathematics class may serve to exemplify this concept.

In the first example, the teacher draws five little circles in one row on the chalkboard and colors the first three white (O) and the other two red (●):

$$\bigcirc \bigcirc \bigcirc \; ● \; ●$$

The production of a collective argumentation begins when the teacher explains that today, these objects will be neither zeros nor "o's" but circles. Shortly thereafter the discourse develops as follows[3]:

1. *T:* Today I want to do something different. I want to write down a number sentence. First I drew five circles....

2. *S:* Plus....

3. *T:* These five circles are three white ones ... plus ... two red ones. This is the number sentence that belongs to this. (The teacher adds a second line to the circles on the chalkboard.)

$$\bigcirc \bigcirc \bigcirc \; ● \; ●$$
$$5 = 3 + 2$$

Repeat together with me.

4. *T, Ss:* Five....

5. *T:* These were the five that I drew for you, folks.

6. *T, Ss:* Equals three plus two.

In this episode the point of the argumentation is related to the introduction of a new way of confirming the correctness of the statement $5 = 3 + 2$: referring to a pictorial representation. Because one can put

---

3. The two episodes stem from observations in a German mathematics class. The presented transcripts are translations by the author. Inevitably this translation shows quite more grammatical consistency that the German original (see also Krummheuer [1989, 1991, 1992]). One should notice here that the two examples are not chosen to show paradigmatically "good" mathematics teaching. Because of the ethnomethodological basis of this chapter, these episodes represent rather everyday situations of mathematics teaching. Therefore, the developed concepts attempt to describe the functionality of *everyday* teaching-learning processes and not so much the aspects of desired future mathematics education.

five circles as three white ones and two red ones, one can conclude that the equality $5 = 3 + 2$ holds. This is not the usual way for children of this age to assume a counting strategy. As Hughes (1986) points out, this relationship is not self-evident for first graders. They have to develop ways of translating from the given pictorial representation to the arithmetic one. This translation is seen here as the underlying aim of the argumentation.

This small episode shows roughly how a format of argumentation has been constituted in the interaction process. By her statement, the teacher initiated the way in which she wants to relate the five circles to the number sentence $5 = 3 + 2$ (lines (1) to (3)). As in a choir, she and the children repeat the verbal version of this sentence (4). Hereby the teacher clarifies once more the intended relationship between the picture and the number sentence on the chalkboard (6). After this, the formatted argumentation of the relationship between the decomposition of five circles in two subsets and an additive number sentence is practiced several times.

The first steps of generating a format of argumentation have been accomplished. In the terminology of a theory of argumentation, one can say that criteria have been set (by the initiative of a teacher) that in a specific classroom context help classify a sequence of statements as a commonly accepted argument. At this point it is not necessary to assume that the children already possess or activate an appropriate mathematical rationale for this format of argumentation.

The second example takes place a few weeks later in the same class. The teacher draws six circles on the chalkboard and writes $6 - 3 =$ below.

$$\bigcirc \; \bigcirc \; \bigcirc \; \bigcirc \; \bigcirc \; \bigcirc$$
$$6 - 3 =$$

1. *T:* So, six minus three, and now I'd like to know from you....

2. *S:* Minus three equals....

3. *T:* How I can show that here (*points at the circles*). Obviously I cannot take away three [from here]. From that magnet board over there I took away three girls or three dwarfs and placed them aside. This does not work here now.

4. *S:* Erase it!

5. *T:* No, I don't want that.... If you erased things in your books, it'd look entirely ugly. There is another possibility.

6. *C:* Cross it out, cross it out (*goes to the chalkboard*).

7. *T:* Yes, exactly. Cross three out, then we know... (*C crosses out three circles*).

   Now, if I cross out these three and take them away, like I wrote in the number sentence, how many do I have left?

8. *C:* Three.

9. *T:* Yes, write it down (*C completes the number sentence*).

   Good job.

$$\bigcirc\ \bigcirc\ \bigcirc\ \emptyset\ \emptyset\ \emptyset$$
$$6 - 3 = 3$$

Here again one can reconstruct how the teacher tries to initiate a certain kind of argumentation that regulates the relation between the six circles and the number sentence "$6 - 3 = 3$."

FORMATS OF ARGUMENTATION AND THE LEARNING OF MATHEMATICS

Often teaching-learning processes can be characterized by a constitutive difference of the interpretation of the situation by the teacher and the students. In such classroom situations, in the process of negotiating meaning, participants tend to present their contributions in a way that shows their argumentative power. This serves two functions for the interaction: it explicates one's own thinking, and it helps to achieve a shared interpretation of the situation. (In ethnomethodological terms, a classroom participant tries to make his or her interactional contributions "accountable"; see, for example, Garfinkel [1967, p. 280 ff.].) Presenting one's own contribution in this way helps make it mutually understandable and gives it a status of intersubjectivity.

These kinds of negotiation of meaning can be based on formats of argumentation. In the process of learning, these formats serve to initiate a structure of argumentation that becomes increasingly predictable and autonomously reaccomplishable by the students. Finally the student can argue according to a format in peer work or individual work without the participation of the teacher.

> At the start, formatting is under the control of the adult. Increasingly, formats become symmetrical and the child can initiate them as readily as the adult. (Bruner 1982, p. 12)

The student develops increasing autonomy in acting according to certain formats of argumentation. On this basis he or she cognitively constructs new meanings for the involved concept, which gradually converge to the officially ratified mathematical meaning (see also Light [1987, p. 56]). Thus, conceptual mathematical learning is based on autonomous processes of argumentation that have been experienced in social situations of formatted collective argumentation. Formats of collective argumentation can be seen as a specific way of assisting the learning of mathematics in the classroom interaction (Bauersfeld 1991; Krummheuer 1995).

The mathematics class represents a social *topos* or *locus* (Perelman and Olbrechts-Tyteca 1969, p. 83; Krummheuer 1995, p. 259) where content-specific formats of argumentation are generated. Learning mathematics is situated in the participation structure of a classroom culture of arguing. Classroom discourse, emerging according to such formats of argumentation, is seen as a social setting of generating the rationality of mathematics. Thus mathematical rationality is seen here as the product of specifically formatted processes of interaction and not as their presupposition.

REFERENCES

Bauersfeld, H. "Integrating Theories for Mathematics Education." In *Proceedings of the Thirteenth Conference of the International Group for the Psychology of Mathematics Education, North American Chapter.* Blacksburg, Va., 1991.

Bauersfeld, H., G. Krummheuer, and J. Voigt. "Interactional Theory of Learning and Teaching Mathematics and Related Microethnographical Studies." In *Foundations and Methodology of the Discipline of Mathematics Education*, edited by H. G. Steiner and H. Vermandel. Antwerp: University of Antwerp, 1985.

Bruner, J. *Acts of Meaning.* Cambridge: Harvard Univerity Press, 1990.

———. *Child's Talk: Learning to Use Language.* Oxford: Oxford University Press, 1983.

———. "The Formats of Language Acquisition." *American Journal of Semiotics* 3 (1982): 1–16.

———. "The Role of Interaction Formats in Language Acquisition." In *Language and Social Situations*, edited by J. P. Forgas, pp. 31–46. New York: Springer-Verlag, 1985.

Cobb, P. "Mathematical Learning and Small-Group Interaction: Four Case Studies." In *The Emergence of Mathematical Meaning: Interaction in Classroom Cultures*, edited by P. Cobb and H. Bauersfeld, pp. 25–130. Hillsdale, N.J.: Lawrence Erlbaum Associates, 1994.

Cobb, P., E. Yackel, and T. Wood. "Young Children's Emotional Acts While Doing Mathematical Problem Solving." In *Affect and Mathematical Problem Solving: A New Perspective*, edited by D. B. McLeod and V. M. Adams, pp. 117–49. New York: Springer-Verlag, 1989.

Forman, E. A. "Discourse, Intersubjectivity, and the Development of Peer Collaboration: A Vygotskian Approach." In *Children's Development within Social Contexts: Metatheoretical, Theoretical, and Methodological Issues*, edited by L. T. Winegar and F. Valsiner, pp. 143–61. Hillsdale, N.J.: Lawrence Erlbaum Associates, 1991.

Garfinkel, H. *Studies in Ethnomethodology*. Englewood Cliffs, N.J.: Prentice Hall, 1967.

Hughes, M. *Children and Number: Difficulties in Learning Mathematics*. Oxford: Basil Blackwell, 1986.

Inhelder, B., and J. Piaget. *Growth of Logical Thinking from Childhood to Adolescence*. New York: Basic Books, 1985.

Keil, F. C. "On the Emergence of Semantic and Conceptual Distinctions." *Journal of Experimental Psychology: General* 112 (1983): 357–85.

———. *Semantic and Conceptual Development: An Ontological Perspective*. Cambridge: Harvard University Press, 1979.

Kopperschmidt, J. *Methodik der Argumentationsanalyse*. Stuttgart Bad Cannstatt: Frommann Holzboog, 1989.

Krummheuer, G. "Argumentations Formate im Mathematikunterricht." In *Interpretative Unterrichtsforschung*, edited by H. Maier and J. Voigt. Cologne: Aulis, 1991.

———. "Die Veranschaulichung als 'formatierte' Argumentation im Mathematikunterricht." *Mathematica Didactica* 12 (1989): 225–43.

———. "The Ethnography of Argumentation." In *The Emergence of Mathematical Meaning: Interaction in Classroom Cultures*, edited by P. Cobb and H. Bauersfeld, pp. 229–70. Hillsdale, N.J.: Lawrence Erlbaum Associates, 1995.

———. *Lernen mit "Format." Elemente einer interaktionistischen Lerntheorie. Diskutiert an Beispielen mathematischen Unterrichts*. Weinheim: Deutscher Studienverlag, 1992.

Krummheuer, G., and E. Yackel. "The Emergence of Mathematical Argumentation in the Small-Group Interaction of Second Graders." In *Proceedings of the Fourteenth Conference of the International Group for the Psychology of Mathematics Education*, vol. 3, pp. 113–20. Oaxtepec, Mexico, 1989.

Light, P. "Taking Roles." In *Making Sense*, edited by J. Bruner and H. Haste. London: Methuen, 1987.

Miller, M. "Argumentation and Cognition." In *Social and Functional Approaches to Language and Thought*, edited by M. Hickmann. San Diego, Calif.: Academic Press, 1987.

————. *Kollektive Lernprozesse. Studien zur Grundlegung einer soziologischen Lerntheorie.* Frankfurt am Main: Suhrkamp, 1986.

Perelman, C., and L. Olbrechts-Tyteca. *The New Rhetoric. A Treatise on Argumentation.* Notre Dame, Ind., and London: University of Notre Dame Press, 1969.

Piaget, J. *The Child´s Conception of Physical Causality.* London: Routledge & Kegan Paul, 1930.

————. *The Construction of Reality in the Child.* New York: Basic Books, 1954.

————. *Judgement and Reasoning in the Child.* London: Routledge & Kegan Paul, 1928.

Rogoff, B. *Apprenticeship in Thinking.* Oxford: Oxford University Press, 1990.

Siegal, M. *Knowing Children: Experiments in Conservation and Cognition.* Hillsdale, N.J.: Lawrence Erlbaum Associates, 1991.

Struve, H. *Grundlagen einer Geometriedidaktik.* Mannheim: BI Wissenschaftsverlag, 1990.

Toulmin, S. E. *The Uses of Argument.* Cambridge: Cambridge University Press, 1969.

Turner, J. H. *A Theory of Social Interaction.* Stanford, Calif.: Stanford University Press, 1988.

Voigt, J. *Interaktionsmuster und Routinen im Mathematikunterricht.* Weinheim: Beltz, 1984.

————. "The Social Constitution of the Mathematical Province—a Microethnographical Study in the Classroom." *Quarterly Newsletter of the Laboratory of Comparative Human Cognition* 11 (1989): 27–34.

————. "Thematic Patterns of Interaction and Sociomathematical Norms." In *The Emergence of Mathematical Meaning: Interaction in Classroom Cultures,* edited by P. Cobb and H. Bauersfeld, pp. 163–202. Hillsdale, N.J.: Lawrence Erlbaum Associates, 1995.

Völzing, P. L. *Kinder argumentieren.* Paderborn: Ferdinand Schönigh, 1981.

Wittgenstein, L. *Tractatus Logico-Philosophicus.* London: Routledge & Kegan Paul, 1963.

14

Teaching without Instruction: The Neo-Socratic Method

Rainer Loska
University of Erlangen-Nürnberg

I N THE history of education, many school reformers have been fascinated by the Socratic method as presented in Plato's dialogues (Plato 1987). The section in *Meno* about the doubling of a square has especially served as a model for a teaching lesson in mathematics. What the core of the Socratic method should be has been the subject of controversy for centuries. Plato characterized Socrates' method as an art of midwifery—the art of assisting at the birth of thoughts, the art of maieutic—that realizes an ideal of teaching: The teacher does not really teach; the student or the partner himself discovers the solution "from his inner self." Socrates said: "Do you observe, Meno, that I am not teaching the slave anything, but only asking him questions?" (*Meno* 82e). Here he postulated a disjunction between teaching something and a nonteaching method that uses questions. Hence the element of question gained a special place in theoretical discussions about education. Since that time, questioning on one side and explicit teaching of the respective subject matter on the other side have been at opposite poles in discussions on teaching methods. But this rigidity has hampered the development of the method to overcome its obvious deficiencies. This chapter introduces the neo-Socratic method, rooted in the same tradition but offering an innovative solution for the problem of maieutic teaching. The next section considers how Socrates conducts the dialogue.

SOCRATES' METHOD: TYPES OF QUESTIONS

Questions are indeed a decisive instrument for Socrates. Moreover, Socrates' role within the dialogue is determined by the use of the question as a tool. If we look at this tool more closely, we find two categories of questions, one of which consists of the *yes-no questions*, for which Socrates' partner's reaction is reduced to agreeing or disagreeing with the content of the question. "Is it not true that ...?" Socrates asks, and his partner answers something like "of course" or "it is so exactly." The questions deliver the decisive thoughts and arguments. In Frege's terminology, the yes-no questions include the entire thought (Frege 1976).

The other, and less frequent, type of question is the *supplementary question*, for instance, "How long is the side of this square?" Whereas the yes-no question corresponds to a proposition, the supplementary question corresponds to a propositional function. Using the instrument of those two types of questions, Socrates is in control of the dialogue. His partner can still influence the course by stating, "I don't understand," which, however, happens rarely. The partner's role consists basically in following the argumentation almost completely delivered by Socrates in the form and in the sequence of questions. Thus Socrates does not fulfill his own maieutic claim.

But the essential fact remains that the partner is asked whether he really agrees to the proposed arguments. He is called on to give his own judgment, and he needs not merely accept a given assertion. The assent of the student transforms the thoughts expressed by propositions into assertions with which he deals.

SOME HISTORICAL ASPECTS

The Meno dialogue has fascinated many people concerned with educational questions from antiquity to the present. We can track the traces of the maieutic idea in the Roman era (e.g., Cicero, Quintilian, and others); in its resumption in the humanistic era (Marsilius Ficino, the translator of Plato's dialogues); and in modern times, starting with Montaigne, an ardent critic of the educational methods of his time, and with his contemporary, Pierre Charron, who explicitly spoke of the Socratic method in his work *De la Sagesse*. In the seventeenth and eighteenth century in Europe, we find the Socratic method booming, especially in the period of the Enlightenment.

However, some confusion existed about what the Socratic method really is. In the ecclesiastic tradition, the method of catechizing was established, and there we find many different types of recitations that claim to be Socratic. Most of those lessons have just two structural features in common: questions of the yes-no type or of the supplementary type and a one-to-one social relation, that is, one person who plays the teacher's part and another person, the student, whose role is to answer. Most lessons start with an oral or written instruction presented by the teacher, without any discussion. Then the teacher asks questions to examine whether the student can memorize the subject matter and has some verbal understanding. This kind of teaching is also called Socratic, although the student's own judgment is not asked for. One can find examples for which the order of the questions is arbitrary and has no logical structure at all. Examples exist in which the teacher asks questions to determine whether the student has reached a somehow deeper understanding and is not only memorizing.

Many models of Socratic lessons dealing with different topics were published. One of the first maieutic dialogues on a mathematical topic was written by Leibniz about 1676 but published only in 1976. An early example of a mathematics lesson can be found in a teaching book for the education of teachers by August Hermann Francke (1971), revised by his collaborator Hieronimus Freyer in 1721. In the nineteenth century, the young mathematician Karl Weierstrass wrote in praise of the Socratic method and thought it the best method for teaching mathematics (1903). At the same time he stated its impracticability, for in the public educational institutions it is not possible for a teacher or a professor to concentrate on one student only. If the method is applied to a group or a class, it will be diluted and will lose its efficiency. Thus Weierstrass's praise is, at the same time, a farewell.

In the development of the Socratic method, attempts have been made to apply the method in the classroom situation without lessening the power of the one-teacher–one-student situation. Thus, in the nineteenth century, Tuiskon Ziller developed his method of disputation (Ziller 1876, p. 139–53), a method based on the idea of teaching indirectly. He consistently tried to induce his students to run a discussion relating themselves to one another. The teacher's new role was to guide the discussion among the students. Ziller used various means, like giving hints, encouraging the students, and confirming or objecting to the statements by his gestures. Ziller himself, as a well-known representative of the Herbartian school, was widely respected. But his method of disputation was hardly acknowledged. In the literature of the end of the nineteenth century, the topic of the Socratic method was not a point of discussion despite the fact that variants of catechizing lessons claiming to be Socratic had been widely practiced.

LEONARD NELSON'S NEO-SOCRATIC METHOD

At the beginning of the twentieth century, a German philosopher, Leonard Nelson, the first in Germany to be a professor of philosophy at a faculty for mathematics and natural sciences, rediscovered and restructured the Socratic method to the neo-Socratic method so it could fulfill its maieutic claim. He was an admirer of David Hilbert in Göttingen and was supported by him in his university career. Nelson criticized the way of teaching philosophy at the universities, in which students of philosophy gained some knowledge of the opinions of philosophers but did not really learn how to philosophize. He formulated the basic problem of maieutic teaching: "How is any instruction, and therefore any teaching, at all possible when every instructive judgment is forbidden?" (Nelson 1949, p. 18). His innovation started with the application of the method to a group of students. He abandoned the one-to-one social structure. But in contrast to his eighteenth-century predecessors, he not only enlarged the number of the students, he redefined the role of the teacher. The most important condition was that the teacher was forbidden to utter his judgment in the subject matter, including the right-wrong evaluation of the students' statements. Thus the teacher was freed from any contribution to the topic. In short, his job was to take care of three things: a genuine mutual understanding among the students, the concentration on the respective question to prevent digression, and the preservation of the good ideas that had come up in the course of the discussion.

The ideas and the argumentation were to come from the students themselves, and it was up to them to arrive at their own conclusions. The teacher was not applying those types of instructive yes-no and supplementary questions that were used by Plato's Socrates. His questions or requests could be, for instance, "Did you understand what she just said?" or "Do you agree with her statement?" or "What does it mean with regard to our question?" These questions were regulating the course of the discussion; let us therefore call them regulating questions. They did not add a new thought, and they did not imply a right-wrong evaluation of the statements of the students. Thus the condition for a genuine maieutic claim was ensured. Socrates had been forced to use argumentation because a teaching discourse between only two partners could not have otherwise developed. His questions had functioned to keep the discourse constantly alive and to come to a result. Only the change to a plurality of participants made it possible to free the teacher from contributing thoughts to the topic. The consequence was a reduction of the teacher's means of intervention. The teacher was forced to keep to himself, which was in contrast to many teachers' habits. All the time, however, he had to be aware of what was happening in the discussion with regard to the students' dealing with the problem. The students' responsibility, in contrast, was to develop ideas and explain them, to develop chains of thoughts. Incidentally, their quantitative share was much higher than usual. The teacher intervened in the following situations:

- He wanted to get information on how far the students could follow the discourse.
- He got the impression that some students did not follow.
- He observed that a contribution was not in direct relation to the respective question.
- He realized that an important new idea had arisen or that a misleading, but strongly stated, opinion was included in a contribution.

The teacher's means to conduct the discussion were the regulating questions. He could also strengthen a contribution by writing it on the chalkboard. Thus the students got no information on whether the teacher agreed or disagreed. They could only conclude that the emphasis meant that the statement should be examined. Its verbal or literal representation helped them concentrate on the question.

AN EXAMPLE OF THE NEO-SOCRATIC METHOD IN A MATHEMATICS LESSON

A brief description of a two-period (about 120 minutes altogether) class held in the eighth grade illustrates the neo-Socratic method. The task in this class was to construct a regular pentagon for which the length of the sides was given. A student teacher gave a short introductory lesson about regular polygons just before the lesson started. The pentagon was chosen as a special polygon to be constructed with compasses, straightedge, and protractor. In the case of the regular pentagon, compasses and straightedge alone are needed. Yet it was not evident how the students could find the solution by themselves if they did not have any notion of, or any experience in, the construction of the medial section that would be indispensable here. For the given problem they could refer to their experience. The students started with the idea to compose the pentagon from five isosceles triangles. One of the students had already determined that the central angle measured 72 degrees, which had to be clarified for the rest of the class. As a first step, students worked out the construction of a regular pentagon, starting with the central angle. All students were able to follow this procedure. The description of the construction was formulated by them: "I drew a circle and a radius from the center. Then I added there four times an angle of 72 degrees with the protractor. I connected the endpoints I had found."

At the beginning of the next period, this text was criticized as insufficient. A student proposed to specify the term *endpoints* and replace it with *points of intersection on the circle*. (This growing need for precision is an interesting phenomenon, which can be frequently observed in neo-Socratic discussions.) After that, the students made three different proposals about how to continue:

1. To generalize the discovered method of construction for any regular polygon

2. To continue to find the solution for the original question—how to construct the pentagon with given length of side

3. To look for another construction without using the circle

All proposals were reasonable. Obviously it was not obligatory to take the direct way. In fact, during the course of the discussion, students may find some special aspects interesting or even exciting and become interested in following them. In this class, proposal 2—to solve the original task—was supported by the majority.

The next idea was to begin with side AB and to erect the perpendicular bisector (fig. 14.1). The center M of the circle had to lie on this line. A student demonstrated on the chalkboard and stopped drawing the perpendicular bisector, assuming the endpoint as the center of the circle. He connected A and B with the thus found point M. When executing this proposal, the students found that M was not determined by an arbitrarily found point M—most of them interpreted M as the point of intersection of the two circle segments when the bisector was constructed. As they continued, they observed that the five congruent tri-

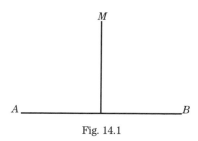

Fig. 14.1

angles did not fit into a pentagon. They realized that the angle in the middle was not automatically 72 degrees. But this step was necessary.

Some further proposals were made. One of them started with the idea to draw an angle of 72 degrees five times around a point M to figure 14.2.

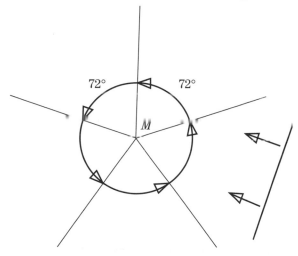

Fig. 14.2

But the students did not follow up this proposal and look for a way to adjust the sides correctly. It was supposed that their experience up to that point was not sufficient to apply the required knowledge without too much effort. The problem of continuing from this point did not arise because another student had found the solution by calculating the value of the angles on the base of the triangle. The sum of three angles in the triangle is 180 degrees. One angle in the center is 72 degrees; the other two are equal, so their measure is 54 degrees. Thus it is possible to construct the triangle by starting with AB and drawing two 54-degree angles, at A and at B. The intersections of the sides determine M, where the angle has consequently to be 72 degrees. To add the congruent triangles five times leads to the regular pentagon with sides of the given length.

The following remark concerns the student whose thoughts went the furthest: he had difficulties in expressing his ideas in a way that could be understood by the others. But another student immediately grasped his ideas and was in a position to express them so clearly that the others could understand them easily.

When performing the construction, the students had other ideas to vary it, for example, to start with AB and to draw angles of 108 degrees at A and B and so on. For this construction, compasses were not needed. Even weak and quiet students were able to follow the procedure. When the students were asked to comment on these lessons, their statements were positive. One of them said: "It is great to be able to think!"

THE STUDENTS' PART

Let us look at the work done by the students. It is their responsibility to explore the whole topic without instructive help from the teacher. This process is closer to problem solving in real life, where a teacher is usually not available. It includes blind alleys and mistakes to be corrected; in short, it involves building and testing assumptions. It is a genuine inquiring process with all its features intact. The students always have to think (which had already been shown in Plato's Socratic dialogues)—first, about whether they understand what is being said and, second, about whether they agree with what is said or object to it or doubt it.

It is very important for the students to learn to observe their own reactions to the ideas expressed. A special kind of metacognition is practiced to attend to the accompanying perceptions during the cognitive process. These perceptions do not include those of emotions because they differ from emotions like fear, anger, and joy. Becoming aware of one's own reactions in the cognitive process leads, for instance, to the well-known thought, "I know that I don't know." This causes the student to think anew about the topic. The teacher's questions, whether all have understood or whether they agree or not to the proposals, serve to initiate reflective thinking. In discussions it happens rather often that the partners can state the fact that they have doubts or that they do not agree, but they cannot say why. It is the task of the teacher, and also of the group, to help them to find out the up to now hidden arguments, ergo, to make them explicit.

We can apply the theory of personal knowledge of Michael Polanyi (Polanyi 1983), who spoke of the tacit dimension when investigating the nature of human knowledge. We can also reconsider the observations of scientists about their way of thinking, but it is also an everyday experience in our own thinking processes. It happens, for instance, that a participant does not agree to a statement. He or she has the perception and can express that the statement is not correct. But he or she is unable to give a proof for this nonacceptance. Some of the various reasons for this follow:

- He or she has a different understanding of the words used by others or of the formulations of the propositions.
- He or she has a prejudice that makes him or her reluctant to accept the statement.
- He or she has the feeling that the argument is inadequate or incomplete.

To examine ideas in a group, students must communicate their thoughts and their perceptions of their reactions in the cognitive process.

THE TEACHER'S PART

It is one of the teacher's tasks to ensure a good atmosphere that allows the students to express their thoughts and perceptions readily. In a social process the class learns to contribute to the process of "making explicit what is implicit," as John Dewey once wrote (Dewey 1986, p. 280), which is a process of clarification. If the students wait for the teacher's judgment instead of relying on their own, this process would be disturbed. If the teacher interferes more and signals his or her judgment even by something as seemingly insignificant as lifting his or her eyebrows, the students are not consistently guided to develop into autonomous thinkers.

In the maieutic discussion, language is important in the thinking process and in the communicative process. To promote mutual understanding it is necessary for the students to make an effort to express themselves clearly. This is not an artificial request made by the teacher because he or she wants it or thinks it necessary. It is a natural request for gaining a better understanding, both for avoiding misunderstandings and for the process of clarifying the ideas. So in a maieutic discussion, an apparently high degree of redundancy can be found. Redundancy is not an absolute category. What is redundant for one student may not be sufficiently informative for another. What is not sufficiently informative at one point may be highly redundant at another. One of the additional aims of maieutic discussion is for students to learn this effect by their own experience. But they are also able to learn that the process of explaining one's ideas to one another can help clarify one's own understanding or lack of understanding.

The students are not left alone. The teacher's work is to support them to fulfill these operations completely and thus prevent them from misunderstanding and from digressing during discussion. A teacher is needed who feels comfortable with the subject matter so that he or she can concentrate fully on the process of the students' dealing with it.

ADVANTAGES, LIMITS, AND PROBLEMS

A maieutic discussion does not include the phenomenon that is known from the analysis of many teaching discussions in the classroom. Heinrich Bauersfeld called it the "expectation of certain answers" (Bauersfeld 1978). If the organization of the lesson plan and the argumentation are more or less linear (i.e., when they are systematic), the teacher is forced to make all the students think his or her way. The discussion is regarded as controllable and able to be fixed in advance: what ideas should emerge, in what order, and at what point of time. In the course of the lesson, the teacher judges more and more the students' contributions by one criterion: do they match his or her expectations according to his or her plan? Thus he or she deprives himself or herself of the flexibility to react to students' ideas. In many instances it can be shown that the teacher does not even recognize them. On the one hand, the teacher ignores or rejects many contributions. On the other hand, his or her interference is growing. It can lead to a kind of catechizing in which he or she embarks on a series of question-answer sequences. But a natural discussion is characterized not by a firm sequence of short steps but by leaps, repetitions, and reconsideration. The neo-Socratic method does not assume that a discussion can be controlled in that way. The maieutic teacher does not try to make sure that certain ideas emerge in a certain order at a certain time. The discussion is open and can lead to different points of the main effort. The teacher should be prepared by his or her analysis of the subject matter. Compared with most other concepts of teaching, a special problem arises. It is not guaranteed that the topics are distributed in the same order as given in the curriculum. A teacher might be confronted with topics not scheduled at all or scheduled for other grades. It may also happen that the discussion on one topic takes much time.

The experience of a teacher who applied the neo-Socratic method constantly in geometry, Rudolf Küchemann (Küchemann 1931), at first confirms that it will often take longer to find the solution of a problem that could be treated with other methods in less time. But because the thinking process goes deeper and deeper, connections between apparently different topics will be realized by the students and they will gain more and more insight. Therefore, an accelerating factor occurs. Küchemann wrote that he never had any problems dealing with all the topics set out by the curriculum.

FINAL REMARKS

The research of the neo-Socratic method has just begun with interpretative case studies. Future research is still left with many questions to be examined:

- What are the effects of the method on the mathematical thinking of the students?
- Will the language of the students become more accurate?
- Are students becoming able to see more connections among the mathematical topics?
- Can metacognitive effects be stated?

Another question is, What topics are especially appropriate for the neo-Socratic method? Geometric topics seem to be the most fitting, perhaps because constructions make examining new aspects of a problem easier. The visualizations immediately show new relations that may help to find solutions, as can be seen in the lessons previously described. Notes describing a neo-Socratic discussion on fractions, for instance, suggest that it could have been helpful if the students had had visual models at their disposal.

In spite of a lot of open questions, we can state that the neo-Socratic method is the first that satisfies by its structure the old maieutic claim presented by Socrates. It is not a method to teach mathematics as a body of knowledge but to teach mathematics as an activity. It is the experience of those who practice it that the neo-Socratic method can be an exciting challenge for thinking, both for the teachers and for the students.

REFERENCES

Bauersfeld, H. "Kommunikationsmuster im Mathematikunterricht—eine Analyse am Beispiel der Handlungsverengung durch Antworterwartung." In *Fallstudien und Analysen zum Mathematikunterricht*, edited by H. Bauersfeld, pp. 158–70. Hannover: Schroedel, 1978.

Dewey, J. *How We Think*. Carbondale, Ill.: Southern Illinois University Press, 1986.

Francke, A. H. *Schriften über Erziehung und Unterricht*. Berlin: Klönne, 1871. (First published in 1721.)

Frege, G. "Der Gedanke." In *Logische Untersuchungen*, pp. 30–53. Göttingen: Vandenhoeck & Ruprecht, 1976.

Küchemann, R. "Erfahrungen an einer Oberrealschule." In *Beiheft 16 zur "Zeitschrift für mathematischen und naturwissenschaftlichen Unterricht"*. Leipzig/Berlin: B. G. Teubner, 1931.

Leibniz, G. W. *Ein Dialog zur Einführung in die Arithmetik und Algebra*. Stuttgart-Bad Cannstatt: Frommann-Holzboog, 1976.

Nelson, L. "Die sokratische Methode." In *Gesammelte Schriften*, Vol. 1. Hamburg: Felix Meiner, 1970. (English translation in L. Nelson, *Socratic Method and Critical Philosophy*. [New Haven: Yale University Press, 1949; first edition in 1927])

Plato. *Meno*. In *The Collected Dialogues of Plato*, edited by E. Hamilton and H. Cairns, Bollingen Series LXXI. Princeton, N.J.: Princeton University Press, 1987.

Polanyi, M. *Personal Knowledge: Towards a Post-Critical Philosophy*. London: Routledge & Kegan Paul, 1983.

Weierstrass, K. *Mathematische Werke*, III. Band. Berlin: Mayer und Müller, 1903.

Ziller, T. *Vorlesungen über Allgemeine Pädagogik*. Leipzig: Heinrich Matthes, 1876.

Part 4

Problems of Communication in Specific Domains of Mathematics

15

The Role of Natural Language in Prealgebraic and Algebraic Thinking

Ferdinando Arzarello

University of Turin

THIS chapter develops a theoretical analysis of the gap that exists between arithmetic and algebraic thinking. It proposes a framework for a better understanding of the nature of the transition from an arithmetic (in some way meaningful) understanding to a (possibly) "truly" symbolic algebraic language, through a syncopate style, that is, when algebraic symbols are used but the stream of reasoning is still sustained by natural language. This terminology was introduced by Nesselmann (1842) in studying ancient eastern algebra; it was adapted to educational research by Harper (1987). The transition referred to is accompanied by processes of evaporation of meaning, which have to be converted into processes of condensation. This, in turn, requires a change of perspective; that is, an epistemological transition from a procedural to a relational perspective. Such processes typically concern the inner world of subjects wherein natural language can play a crucial role. The major claim made in this chapter is that the habit of using language in problem solving helps condensation in the long run because it helps students make explicit their inner language.

The role of the verbal code, and specifically that of natural language, is crucial for developing an algebraic way of thinking in solving problems at the secondary school level. However, it must be considered carefully because its correlation with students' performance is complex

I wish to thank Luciana Bazzini (University of Pavia), Paolo Boero (University of Genoa), and Giampaolo Chiappini (IMA, CNR Genoa) for their help in writing this chapter.

and can easily be misinterpreted in everyday teaching. This chapter discusses some of the major points concerning this problem. Of course, it refers to the existing literature, but it also takes into account the recent research of the author on prealgebraic and algebraic problem solving.

Approximately 200 students from 9 to 16 years of age had been involved in the research. They were tested from time to time, and some had systematically participated in sessions of problem solving. Moreover, the author's undergraduate mathematics students at Turin University (80 each year) had been observed as, while training as future teachers, they attended courses and seminars on problem solving and on the teaching of algebra.

This chapter comprises three parts. In the first part, the functions of verbal code in algebraic problem solving are sketched and their positive and negative aspects are discussed; in the second, the dialectic between algebraic and verbal code is analyzed and the role of language in producing formulas as effective thinking tools, and not only as passive knowledge containers, is sketched; in the third part, this dialectic is scrutinized when students face algebraic problems in computer science environments. Hence the focus is on the problem of communicating mathematics (i.e., algebra) to students and by students and on the question of why they are so often speechless mathematically. The dialectic between the inner language and external representations suggests some possible teaching interventions, which are discussed throughout the chapter. Similar problems are discussed from another point of view by MacGregor (in chapter 16 of this volume), who is concerned with the gap between intuition and thought procedures in students' interpretation of equations.

POSITIVE AND NEGATIVE ASPECTS OF INTERACTION BETWEEN THE VERBAL AND THE ALGEBRAIC CODE

The verbal code in problem solving has at least the following main functions:

1. While posing a problem, as a tool to sustain semantic references, which help students enter into the problematic situation, and as a tool to activate, particularly at the level of inner language, those processes that allow them to recall similar problems and (solution) strategies

2. While solving a problem, particularly at the level of inner language, as a tool to construct and control solution strategies. Here, the verbal code has two specific functions. First, it allows the gradual translating into words of one's intellectual activity (thus sustaining the solution strategy); second, it allows one to "input" and "output" the solution process of the problem (i.e., to look at it from an outer point of view, to compare different processes, to rebuild the starting problematic situation, etc.).

3. To communicate knowledge and solutions of problems

4. When constructing mathematical models, as a tool to select relevant features of the phenomenon, to relate them in a qualitative fashion, to refer to analogous situations, and to adapt known models to the new situation (see Arzarello [1992, appendix A])

5. To clarify and properly distinguish concepts used to solve the problem, both at an individual (as inner language) and at a social level (as outer language)

6. To reflect on the knowledge that one constructs, referring to the used methods and the chosen representations; that is, again to help in the input and output of scientific knowledge

Studying students with verbal difficulties can be interesting. It can be verified that their progress in the linguistic field corresponds to progress in the foregoing functions while solving problems. Moreover, many students who do not use verbal language in solving problems—because of the didactic choices of their teachers—reveal low performance in grasping the problematic situation, in building and monitoring solution strategies, in explicating their reasoning, and in reflecting on the results obtained and the chosen solution procedures.

It is worthwhile to discuss the interrelationship between verbal competencies and such intellectual functions as planning or reflecting. It has been observed that the use of verbal code is widespread in solving nonroutine problems. In fact, many students, from middle school to the university, solve algebraic problems more in a syncopate style than in a symbolic one (Harper 1987). Students of 16 years of age do not yet spontaneously use the algebraic code while solving simple algebraic questions, even if this avoidance causes them trouble and long detours. Their preferred models are still arithmetic, which may cause conflicts with the algebraic way of thinking. The only apparent way for them to master the situation semantically often consists in using a syncopate language that allows them to cope with the situation while avoiding conflicts between the problem and the conceptual model they are using. The point is that algebra is understood only in abstraction, as a formal and general tool, but is not concretely used as a method of justification and generalization in specific situations. Generalization as an effective and operative method is used with great difficulty by students while solving algebraic problems. Even if they seem to manage to deal with the formalism, they do not use algebra spontaneously in concrete problematic situations and seem to live more comfortably with its substitutes, namely, those arithmetic models in which the meaning of the relations involved is expressed more directly by the syncopate language of arithmetic. In fact, students are often able to generalize a problematic situation in the form of an algebraic rule, but they describe it in a syncopate arithmetic way. It is too costly to use symbols in a more abstract manner, that is, not as empirical objects, qualities, and procedures of the real world but as signs that incorporate relations and structural connections between objects and qualities. Hence, they control the situation semantically for a long time, postponing the algebraic reasoning until the end; for most of them, only a detour through the arithmetic syncopate language and a procedural style (see below and Sfard [1991] for this concept) can settle the question.

Students generally learn at school how to solve specific problems with algebra (e.g., using unknowns), but they meet major difficulties when required to use it as a symbolism to express general solutions: to discover, generalize, and prove laws behind numerical relations. The main difficulty lies in integrating formal (algebraic) algorithms with arithmetic models. Also here, as in the arithmetic case, one is faced with the phenomenon of the opposition between the model used and the situation to interpret, so that to overcome difficulties, one must adapt the model to a different situation and get a new model. But this phenomenon has a different conceptual character. The opposition is not between the problem and a model that is not adequate. The point now is that conceptual arithmetic models are adequate to attack the situation, but only from an arithmetic point of view. The opposition is between the very way in which one has the habit of using the model and the necessity of looking at it in a different, more abstract fashion. Conceptual models must be transformed, but a partial modification is not enough. They must be inserted into a more general framework. In fact, many students who cannot solve a problem do not see things, even if they are under their eyes. The difficulty is to look at arithmetic things with a new (algebraic) eye, so to speak. Students must learn to see and speak according to a new code, which means a shift in abstraction. Also, the habit of using the verbal code in problem-solving sessions seems to be essential to make learning such a new language possible for them.

This point illustrates the weakness of the typical controversies in proposals for teaching and learning algebra that concern the very way in which the algebraic signs should be presented to the students, for example, concrete letters versus abstract variables. (For more on the question of whether natural language can be a tool in supporting students' algebraic thinking, see also the contributions of Pirie (chapter 1) and MacGregor (chapter 16) in this volume.) The essential point is the context in which students are taught to use and manage signs. If the focus is mainly on manipulative aspects and, moreover, if the manipulations taught all have the same features (e.g., simplifying algebraic expressions or factoring polynomials, which are taught in Italy to 14- to 16-year-old students), students might naturally interpret the meaning of letters as "pure" signs, forgetting the structural relations that they can embody. Unfortunately, this situation can be the root of a vicious circle by which the teacher and the student use the same words (e.g., *variable* or *letter*) or the same representations (in an algorithm or a diagram or a graph) with different meanings: The teacher's interpretation and representation may not be like those of the students. Words, sentences, formulas, models, diagrams, and colors are not "transparent" and may not be helpful to students. The disparity between teachers' explanations and students' understanding is one of the most typical stumbling blocks in teaching algebra. There starts a "comedy of errors" that can have dramatic consequences.

Among the main novelties of the algebraic code, when compared with that of the natural language (e.g., when it is used by students to explain their solutions in arithmetic problems), one finds the so-called ideographic and transformative functions, which are missing or obtained in essentially different ways in the verbal code. But these functions are costly for the students, who prefer to use more stable models, even if inadequate, namely, to supply such algebraic functions with extra linguistic properties. The language of students faced with algebraic word problems is full of linkages to different aspects of the real space-time situation in which they act while producing their own solution of the problem. The protocols contain an easy-to-find melting pot whose ingredients are mathematical, extramathematical, extralinguistic, and so on. Typically, mathematical objects are referred to by means of the subject's actions (e.g., processes of calculation made by the subject or by somebody else); algebraic laws are put into the flowing stream of time. All this makes it easier for students to remember the meaning of the formal things about which they are speaking and to control the situation semantically. It enhances their use of the often stable arithmetic conceptual models, even if this use has a local character only and is costly in terms of memory. This style decreases very slowly in the course of the school years. A relatively low difference has been observed by the author between 10-year-old children and mathematics university students.

By means of natural language, in a syncopate manner, students can express the meaning of mathematical objects, referring to the subject's actions, the very processes of their constructions and generations, and other extramathematical information about them. This is a major source of obstacles to ideographic and transformative functions; in fact, these may cause an evaporation of extramathematical data and, consequently, a possible dramatic loss of meaning.

Before introducing this point, which is at the heart of the next part of the chapter, let us conclude this first section by pointing out some limits of the verbal code when it is contrasted with the algebraic one.

It has already been mentioned that some functions are specific to the algebraic code (i.e., ideographic and transformative functions) and are otherwise surrogated or missing in the verbal code. Another different function, also missing in the verbal code, is the shorthand function, which may cause certain mistakes, but less dramatic ones, however, when they are compared with the epistemic obstacles rooted in the other two. Another interesting limitation of the verbal code is that it can properly express qualitative relations among variables but is less adequate for expressing quantitative relations. For example, additive situations enjoy descriptions in a highly sophisticated verbal code, but the code becomes inadequate and misleading for multiplicative situations and is even worse for other arithmetical relations. In the long run it becomes almost useless. This is a good point of discussion. Anthropological and psychological research studies are needed to clarify the genesis of obstacles to the proper use of models incorporated into the verbal code.

THE ROLE OF LANGUAGE IN THE TRANSITION TO THE ALGEBRAIC CODE

Evaporation—that is, the dramatic loss of the meaning of symbols met by many students when they abandon the syncopate style in algebraic problem solving—is one of the main obstacles to developing an algebraic way of thinking (concrete examples can be found in Arzarello [1992]). There are two main difficulties:

1. To put concrete (i.e., numerical) relations into a general formula
2. To transform the formula properly to get the solution of the problem

At this delicate point, the role of language is essential, particularly insofar as the dialectic between inner and outer language bypasses evaporation, provoking instead a new and positive phenomenon, called condensation, which is essential for the development of the functions of the algebraic code. Condensation is apparently a sudden phenomenon—it happens on the spot—but it can be grasped properly if one does not see the learning and teaching of algebra as a sequence of single acts of language but as a sort of stream of thought. This breaks dramatically with the arithmetic way of thinking with which students have been acquainted since elementary school. Other authors, like Chevallard (1985, 1989, 1989–90), have discussed this problem from an epistemological point of view; here it is investigated from a linguistic point of view.

Observing students who solve algebraic problems effectively is intriguing because of the (apparently) chaotic numerical collecting of data they do first and the subsequent sudden transition to ideographic and transformative functions. Interviews with all students have eventually clarified some points, which are discussed next.

Many solutions of nonroutine algebraic problems live dialectically in a double polarity. On the one hand is the subject who solves the problem, with his or her actions in the flow of time; the algebraic code is interpreted essentially in a procedural manner. Hence arithmetic models can be easily adapted and integrated into this way of thinking. On the other hand, the algebraic code must be interpreted in an absolute way, independently from the actions of anybody. It is a contemplation of relations and laws sub specie aeternitatis: neither procedures nor products of actions are involved. In fact, only the abstract-relational aspect remains and the sense of the letters becomes very abstract, insofar as it encompasses mutual functional relationships of variables (and possibly of parameters).

Procedural polarity is asymmetric, has a privileged direction, and is controlled by means of tense adverbs and prepositions; its logic is the "logic of when." Relational polarity is ruled by logical and equivalence laws, which are typically symmetric; its logic is the "logic of if and only if." The former allows for a strict and direct semantic control.

Extralinguistic facts steadily link the speech to the subject's actions in the flowing stream of time. The latter, on the contrary, is typically symbolic. Concrete meaning has evaporated and condensed into symbolic objects, which do have their own semantic, albeit a very formal one, since they depend on the condensation of previous extralinguistic facts in a symbolic, ideographic fashion (it is a sort of second-order way of looking at things, namely, using relations of relations).

In the procedural polarity, the dominating epistemological style is arithmetical. Calculations are performed as soon as possible. Every (syntactical) term is developed at once, until it is calculated, that is, transformed into an irreducible one (number written in a canonical form). The epistemological model compels one to reduce the formal complexity of terms, without caring about their numerical complexity, namely, the number of digits required to write them, for example, in base ten. On the contrary, the epistemological style of algebra sometimes requires one to increase the formal complexity of a term, at least locally, which possibly prevents the growth of its numerical complexity.

Condensation is produced by the transition from procedural to relational polarity. Often it is helped by hypothetical reasoning and experiments made by students in a mental space, a sort of fictitious real world they themselves create. Its reality depends on the situation (namely, the problem, the students, the negotiation in the class, etc.). It gives real feedback to the students who have built it and allows them to activate their own conceptual models. The creation of such mental spaces usually helps students avoid stumbling blocks (i.e., evaporation); in other words, it facilitates all those flowing processes, typical of problem solving, by which data can be selected, processed according to one's conceptual models, and integrated into new pieces of knowledge. The most difficult point is that mental spaces are active if (and generally only if) problems are approached from the procedural polarity (with a rich numerical base), but manipulations in such spaces can be done effectively only by destroying extramathematical tracks, that is, by eliminating this very polarity (evaporation) that is the semantic base of the syncopate algebra. Sometimes the main product of this process is condensation. Using the symbolic code, one can write concisely and expressively the amount of information of a term, whose complexity cannot be easily ruled with the syncopate language of arithmetic. Typical products of condensation are global general formulas, which embody the very nature and mutual relationships of the variables involved. Condensation deeply marks the passing from a procedural moment to a more abstract and relational one; it appears at once in the solution strategies of students and, as with all processes that happen on the spot, is very difficult to analyze because attention is focused on single acts of language that reveal only a sudden change. For example, in all examined cases, most students who work successfully on problems and solve them by passing suddenly from procedural to relational polarity are not able to explain what happened, even when explicitly asked. The only positive observed fact is that condensation is marked by a massive use of all functions of language listed in the first part, whereas evaporation is not. Moreover, in all control classes, where the teaching style did not encourage the use of natural language in problem solving, only students with high verbal performances got high scores in algebraic problem solving as well. On the contrary, in classes where teachers develop in their students the habit of using the verbal (spoken and written) code while solving algebraic problems, students of average verbal and mathematical abilities also get good scores.

The claim according to which algebraic thinking marks an epistemological rupture between thinking by single acts of language and thinking by a stream of thought is thus confirmed from a didactic point of view. At this point, a plausible hypothesis is that the habit of using language in problem solving helps condensations in the long run because it helps students make explicit their inner language; it seems likely that the stream of thought, necessary to develop algebraic thinking, is deeply related with it.

In such a way, formulas are not obtained as critically as passive knowledge containers but are constructed in the mental space of ideal experiments and become real thinking tools, which do not close the problematic situation and provoke the loss of semantic control in students. In this sense, the habit of making students explain their intellectual activities becomes essential for allowing them to produce condensations—and not only evaporations.

Inner language is basic also for the transformative function of the algebraic code. In fact, it directs the process of transformation by means of anticipating abilities; in problem solving, choosing the correct transformation may not happen automatically (e.g., it may not be a simplification!). There is a sort of dialectic interplay between the inner world of the student and the external representation of a formula that he or she is manipulating. In fact, the student can anticipate a possible shape of the formula, or the very shape of the formula might suggest some manipulating strategy to him or her. It is necessary to study such processes, deeply connected with the inner language, in which students are requested to anticipate (the effect of transformations) and to grasp globally (some portion of) the initial formula to get some insights on its different possible future configurations. Thus an important problem in the didactics of algebraic language is to orchestrate situations in which such processes of internalization can become objects of teaching and of intervention by the teacher.

THE ROLE OF LANGUAGE IN ANTICIPATING THOUGHT

To attain an algebraic way of thinking, the habit of elaborating generic examples becomes crucial, namely, to neglect all that is not general in the specific problem, to look at letters in a functional way, to interiorize invariance, and to make mental experiments that work concretely in an ideal space without any concrete example of data. The role played by the so-called anticipating thought in such processes is also crucial. Sometimes it operates on the basis of rich numerical tables constructed by students, possibly by using the computer. Sometimes anticipating thought is a sort of hypothetical reasoning made in a mental space that has spatial, temporal, and logical features. Mental space is a sort of imagined real environment. It has some connections with the kind of ideal environments where philosophers and scientists like Galileo, Kant, Faraday, and Einstein made their well-known mental experiments (which is an interesting epistemological point). As a typical metacognitive skill, it is a long-time process, which can be developed by means of a cognitive apprenticeship (see Arzarello et al. [1993], and the examples infra), where the teacher encourages thinking as a stream and not as single acts of language. This encouragement can take the form of scaffolding (Bruner 1985, p. 29) the task of the students, a strategy that reduces the number of the degrees of freedom that they have to manage. Another way is to have the students contrast and discuss their strategies of solution, treating this as their main task in an algebra class.

Hypothetical reasoning is related to problem-solving activities in different important moments and entails the verbal functions discussed in the first part. It is essential both in general planning, when students elaborate possible solution strategies, and in controlling, contrasting, and comparing them with other models.

A second aspect of anticipating thinking as hypothetical reasoning is shown by students faced with numerical tables that they themselves have generated, possibly by using a computer. Also, in these instances, students who can overcome obstacles do many mental experiments, but in contrast to the previous case, in which hypothetical reasoning had mainly inner features, the interaction this time is with external representations. That is, their processes are based on—

- observation of arithmetical patterns and relations between them in a table of numerical values—students do experiments with numbers;
- hypothetical thinking in a mental space that is an ideal extension of the table.

In both situations, most of the verbal functions listed in the first part are very active, particularly those concerning inner language. The table provokes anticipation. It becomes an ideal object, in the sense that it is the starting point for arithmetic experiments that soon become purely mental and produce hypothetical reasoning and validations. In short, hypothetical reasoning is a particular form of anticipating thinking, which can be observed in problem solving, particularly when students are faced with obstacles.

Anticipating thinking appears also in other aspects of problem solving; for example, it has an important role when pupils program a computer. In this example, the crucial problem for them is to put together, in a systematic way, different instructions and commands—to get sequences of commands recognizable as consistent with the global hypotheses formulated in advance—while constructing the solution strategy. Since hypothetical reasoning performances generally indicate that the learning situation has a proximal (Vygotsky 1978) character, they may stimulate a novice's transition from pen-and-paper environments to programming ones. Such situations reveal themselves as particularly rich and intriguing when main algebraic conflicts are dealt with (some examples are discussed in Arzarello et al. [1993]).

FINAL REMARKS

One of the main problems in teaching algebra (and most of mathematics) is a communication problem. The relationship between signs and their mathematical meanings may be confused for many students who attach only formal and procedural features to the former but who use the same words as their teachers, albeit with different meanings, for representing the situation. This is a general source of difficulties in creating good learning situations. Two metaphors, namely, the "theater metaphor" and the "joke metaphor," illustrate this didactic weakness very well. In fact, interaction in the classroom is like that in a theater where actors feature some characters. Students and teachers are both actors and spectators and must act in accordance with their different roles.

This metaphor brings forth an important question. Teachers cannot give the meaning directly; they must "act" it. How? Understanding a mathematical statement or situation is like understanding a joke. A joke does not put its point up front in an explicit way; neither do mathematical statements about their meanings. To understand a joke, or mathematics, one is expected to connect and select the right things to laugh at or make sense of; otherwise, the joke is missed—or the meaning of the mathematics is lost. The question is, What sort of theater can teachers plan in order that their students laugh at the right moment while doing algebra? This metaphor connects learning mathematics with living in a culture. If you are not a member of the culture, or if you do not know the culture, you will not understand what makes the people laugh.

REFERENCES

Arzarello, F. "Pre-Algebraic Problem Solving." In *Mathematical Problem Solving Research*, edited by J. P. Mendes da Ponte, J. F. Matos, J. Matos, and D. Fernandez. NATO ASI Series. Berlin: Springer, 1992.

Arzarello, F., G. P. Chiappini, E. Lemut, N. Malara, and M. Pellerey. "Learning to Program as a Cognitive Apprenticeship through Conflicts." In *Cognitive Models and Intelligent Environment for Learning Programming*, edited by E. Lemut, B. Du Boulay, and G. Dettori. NATO ASI Series. Berlin: Springer, 1993.

Bruner, J. "Vygotsky: A Historical and Conceptual Perspective." In *Culture, Communication and Cognition: Vygotskian Perspectives*, edited by J. V. Wertsch. Cambridge: Cambridge University Press, 1985.

Chevallard, Y. "Le passage de l'arithmétique à l'algèbre dans l'enseignement des mathématiques au collège." *Petit x* 5 (1985): 51–94.

———. "Le passage de l'arithmétique à l'algèbre dans l'enseignement des mathématiques au collège. Deuxième Partie. Perspectives curriculaires: la notion de modélisation." *Petit x* 19 (1989): 43–72.

———. "Le passage de l'arithmétique à l'algèbre dans l'enseignement des mathématiques au collège. Troisième Partie. Voies d'attaque et problèmes didactiques." *Petit x* 23 (1989–90): 5–38.

Harper, E. "Ghosts of Diophantus." *Educational Studies in Mathematics* 18 (1987): 75–90.

Nesselmann, G. H. *Versuch einer Kritischen Geschichte der Algebra*. Berlin: Verlag von G. Reimer, 1842.

Sfard, A. "On the Dual Nature of Mathematical Conceptions: Reflections on Processes and Objects as Different Sides of the Same Coin." *Educational Studies in Mathematics* 22 (1991): 1–36.

Vygotsky, L. S. *Mind in Society: The Development of Higher Psychological Processes.* Cambridge: Harvard University Press, 1978.

16

How Students Interpret Equations: Intuition versus Taught Procedures

Mollie MacGregor
University of Melbourne

EACHERS try to make the concept of equation accessible and easy for beginners by relating it to familiar experiences. Approaches commonly used in Australian secondary schools at the present time are based on four metaphorical models. These are *(a)* a story about a number, *(b)* a function machine, *(c)* a recipe, and *(d)* a balance. The first section of this chapter outlines how these metaphors are used by teachers and why they should be helpful aids to learning. The second section presents evidence that many students think about equations in intuitive ways that do not conform to any of the metaphors. Finally, a theory of intuitive cognitive models of comprehension to account for the disparity between teachers' explanations and students' understanding is presented. The theory explains why certain very simple mathematical relationships are easy to understand when expressed verbally but are hard to represent in algebraic form.

FOUR METAPHORS FOR EQUATIONS

Metaphors play an important role in most communication and comprehension. They enable us to understand one kind of experience or relationship in terms of another that is more directly known. In particular, they enable us to grasp abstract concepts in terms of concrete experiences. According to Lakoff and Johnson (1980), human reasoning may necessarily be embedded in images of events and situations in the physical world. If we did not imagine metaphorical models, abstract reasoning might not be possible at all. Pimm (1981) has shown that not only in ordinary language but also in mathematics, metaphor is essential for the expression of meaning.

The abstract notion of equation is usually introduced in Australian schools by means of one or more of the following metaphors.

The "story about a number" metaphor

An equation can be interpreted as a story about what happens to a number. For example,

$$(4[x - 1] - 2) : 2 = 9$$

may be read as follows:

> There is a number x. You take away 1, then multiply by 4, then take away 2, and then divide by 2. The answer is 9. What was the starting number?

Students draw a diagram of what happens (fig. 16.1).

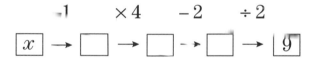

Fig. 16.1

They are shown how to trace the story backward from the answer and write numbers in the empty boxes. They say, for example, "What was the number before we divided by 2? It was 18," then "What was the number before we took 2 away? It was 20," and so on. There is no need for the student to operate on an unknown quantity, since each stage of the method of solution is an arithmetic calculation. This approach to equation solving avoids algebraic processes but has proved to be very successful in enabling beginners to solve equations that have one term, a constant, on the right-hand side. However, it reinforces the incorrect belief that the equals sign has to be always interpreted as separating a procedure on the left from an answer on the right.

The "function machine" metaphor

An equation relating two variables may be interpreted as a rule for a function machine. Like the story metaphor, the function machine represents an equation as a description of what happens to a number.

Students are given tables and pictures, such as those in figure 16.2, to help them understand the notion of a rule that operates on an input number and changes it to the corresponding output number. They are told that each input produces a unique output.

In	Out
2	7
4	11
5	13
10	23
17	37

Fig. 16.2. Function machine

When students are given a set of input and output numbers (usually in table form, as illustrated here), they discover the hidden rule, express it in words, and write it as an equation that relates two variables. The rule for the table in figure 16.2, for example, might be expressed verbally as "The machine doubles the input number and adds 3" and written as $x \times 2 + 3 = y$.

The "recipe" metaphor

Before they start learning algebra, students are already familiar with several formulas, such as those taught for calculating areas and perimeters of geometric figures (e.g., Area $= l \times w$). However, it is unlikely that they see them as examples of algebraic equations. The recipe metaphor is intended to help students link the concepts of recipe, formula, and equation. A recipe can be written like an equation, for example,

1 jug milk = 2 cups milk powder + 8 cups water.

This recipe tells you how to make one jug of milk. Similarly, a formula such as $y = 2(a + b)$ may be seen as a recipe or instruction for working out the value of y. Students easily understand that if they are told the values of a and b, they can work out the value of y. The recipe model gives students practice in substituting values for variables in a formula and calculating. It also shows them that an equation can have an answer on the left-hand side and a procedure on the right.

The "balance" metaphor

In this model, the equation is no longer seen as a description of a rule or procedure. The algebraic terms and numerals are represented as objects in the pans of a balance. For example, the equation $2x + 1 = x + 6$ could be represented as shown in figure 16.3.

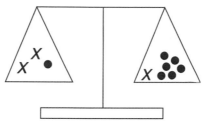

Fig 16.3. Balance

The teacher shows how objects can be added or taken away, or an operation carried out on both sides, which keeps the left and right pans of the balance in equilibrium, until nothing is left except the "unknown" on one side and a numeral on the other.

From a Piagetian point of view, the balance metaphor is appropriate only for students who have reached a formal-operational level of cognitive development. If this is the case, then it is not a useful model for many students in the lower secondary years. However, it is often presented to these students soon after one of the procedural models.

INVESTIGATING STUDENTS' UNDERSTANDING OF EQUATIONS

Teachers assume that students will transfer the structural features of a metaphorical model to the new situation of an equation. However, our research (MacGregor and Stacey 1996) shows that although students learn and remember certain procedures for solving (e.g., reading a number story in reverse or doing the same thing to both sides), they do not necessarily grasp the notion of equivalence between left and right sides. Consequently they have difficulty in constructing equations to represent relationships and in seeing the link between a problem situation and the equation that describes its structure.

To demonstrate students' difficulties, responses to test items from a large representative sample of students aged 12 through 16 are presented and discussed. The youngest students in the sample were in their first year of learning algebra and the oldest were in their fourth year. The items (see fig. 16.4) were included in tests of simple arithmetic and algebra and used in seventeen secondary schools.

---

1. We are told that a and b are numbers and that $a = 28 + b$.
   Which statement must be true?
         (i) a is larger than b.
         (ii) b is larger than a.
         (iii) You can't tell which number is larger.
         (iv) a is equal to 28.

2. I have m dollars and you have k dollars.
   I have \$6 more than you. Which equation must be true?
      $6k = m$ $6m = k$ $k + 6 = m$ $m + 6 = k$ $6 - m = k$

3. s and t are numbers. s is eight more than t.
   Write an equation showing the relation between s and t.

4. The number y is eight times the number z.
   Write this information in mathematical symbols.

Fig. 16.4. Test items

As shown in table 16.1, one-quarter or more of students at all levels did not choose the correct alternative for item 1. Most of the students who were wrong chose "You can't tell which is larger." In some classes this response was chosen by more than 20 percent of students. They associated the algebraic letters with unknown numbers and did not see that the equation offers any information about them.

Table 16.1
Percent of Students Selecting Each of Four Choices in Item 1

		Correct	Incorrect		
Year	n	"a larger"	"b larger"	"can't tell"	"$a = 28$"
7	258	64%	7%	19%	10%
8	253	76%	6%	14%	4%
9	57	66%	4%	23%	7%
10	109	70%	2%	22%	6%

Note: Some students did not respond to this item, although they had plenty of time to do so. Data in the table refer only to the students who did respond.

Item 2 was considerably harder than item 1. Success rates at any level did not exceed approximately 50 percent, as shown in table 16.2. About one-third of the students chose a reversed equation, that is, an equation in which the numeral (6 in this example) is incorrectly associated with the larger variable (m in this example). As may be seen in table 16.2, there are two forms of reversed equations. The most common was

$m + 6 = k$ instead of the correct $k + 6 = m$. The other form was $6m = k$, where $6m$ denotes $6 + m$ or some nonspecific association of 6 and m. An explanation of the reversal error is proposed later in this chapter.

Table 16.2
Percent of Students Selecting Each of Five Choices in Item 2

		Correct		Incorrect			
Year	n	$k + 6 = m$	$6k = m$	$6 - m = k$	$6m = k$	$m + 6 = k$	Omit
7	180	39%	9%	13%	12%	18%	9%
8	215	53%	3%	15%	6%	21%	2%
9	66	47%	2%	12%	6%	27%	6%
10	56	39%	7%	14%	9%	29%	2%

Items 3 and 4 were used in tests for a representative sample of 280 students in their third year of algebra learning. These items proved to be much more difficult than expected. For item 3, only 27 percent wrote a correct equation. Reversal errors, in which the numeral 8 was associated with the larger variable s, were again common, occurring in about one-third of responses.

On item 4, 35 percent of the same sample were correct. As with the previous two items, reversal ($8y = z$) was a frequent cause of error accounting for more than one-third of responses. It is interesting to note that a direct translation from words ("y is eight times z") to mathematical symbols ($y = 8 \times z$, or $y = 8z$), which would have been effective, was not used by most students.

The results for the four items indicate that a considerable number of the students, in spite of two or even three years of learning algebra, did not know how to interpret an equation such as $a = 28 + b$, and a large percentage did not recognize and could not write simple linear equations relating two variables. It appears that the metaphorical models of equations taught in school have not helped them deal with these tasks.

THEORY OF INTUITIVE COGNITIVE MODELS

Kieran (1988) has shown that an important cause of trouble in writing equations is the failure to recognize that the two sides of an equation are equal quantities. This failure seems the most likely cause of errors in item 1. Some students may have interpreted $a = 28 + b$ as "a equals 28, then add some unknown number b." Others may have thought that b was larger, and approximately 20 percent overall could not tell which was larger. Failure in the other three items can be explained by assuming that students based their thinking on conceptual models that did not represent conditions of equality and were not linked with the taught metaphors.

It has often been suggested that students' errors in writing equations are caused by the automatic, intuitive use of natural-language processing rules. These rules, which operate without any conscious intervention or awareness, enable meanings to be constructed in response to spoken and written language. Certain linguistic theories (see, for example, Stassen [1985]) suggest that the rules produce an abstract encoding or a spatial model of a situation and that this model supports comprehension and reasoning. Empirical studies by Osgood (1971) support the existence of a nonlinguistic cognitive system in which comprehension takes place. In this system, the fundamental relationships of contrast, association, and comparison are coded. Johnson-Laird (1983) has used psychological evidence to develop a similar theory, beginning with the notion of a cognitive model that represents semantic features of a situation. Such models represent perceived relationships and attributes in a state of affairs, but they do not necessarily conform to its logical structure. It is significant for a theory of mathematical comprehension that the notion of equality as a link between two elements (a crucial concept for algebra) is not one of the fundamental relationships of cognition proposed either by psycholinguists or by psychologists.

If we assume that such semantic cognitive models underlie comprehension, then the cause of errors in items 2, 3, and 4 is clear. Students were trying to use mathematical notation to express a configuration of mental entities. A student reading item 2, for example, constructs, without conscious intervention, a conceptual model of two entities: one large ("my money") and one small ("your money"). The model supports comprehension and enables inferences to be made (e.g., Who has the most money? How much more? Who has less?). However, the form of the model does not correspond to the syntactic structure of a verbal statement of equality or to an algebraic equation. Students who looked for an equation that matched their model chose the alternatives in which "six dollars more" was associated with the larger amount of money. Similarly in items 3 and 4 they attempted to represent their conceptual models of unequal quantities, incorrectly linking the numeral with the larger variable.

An alternative model that is likely to cause difficulty in the interpretation of an equation is a model of events based on the story metaphor of numbers changing their values when something is done to them. It is possible that errors in items 3 and 4 are derived from such a model. For example, in item 4, since y is eight times as large as another number, there is an implication that it has been multiplied by 8 and that this event is part of the story, which leads to an equation containing the term $y \times 8$ or $8y$. Evidence for such an interpretation was found by Fischer (1988) in discussion with students. Given the equation $x = 3y$, where x and y represent the salaries of two people, students explained that y has been multiplied by 3, so now the two people get the same salary. They seemed to be thinking that "the value of y is now three times its former value, and it equals x."

Teachers find it hard to understand why so many students do not know how to interpret simple equations or how to express a relationship between two quantities as an equation. It appears that the metaphorical models they are taught do not necessarily help them. Instead, as reported in MacGregor and Stacey (1993), many students rely on intuitive conceptual representations of associated or compared quantities, or representations of events, but not of equal expressions. They try to record their conceptual representations on paper by using mathematical symbols.

Students are much more successful in solving simple equations than in interpreting or formulating them. The story and balance metaphors in particular teach students procedures they can use to get a solution. However, it appears that structural features of the metaphorical models, that is, in what ways they represent relations between quantities, are not so easily conceptualized. What students remember about the metaphors are procedures to use in certain contexts, that is, how to get the answer by working backward, how to construct a formula from a table, how to substitute numbers in a formula, and how to do the same thing on both sides of the equals sign. They have not constructed for themselves an understanding of the relationships between the quantities represented in an equation.

It is essential that students learn how to formulate and interpret simple equations as well as how to solve them. All the metaphors are useful as a first introduction to the concept of an equation, and students should be exposed to them all. Each of the metaphors should help students understand particular aspects of the concept of equation and what information an equation can provide. However, each has its limitations. It may focus attention on some feature of the equation concept while hiding other aspects and may impose certain constraints on the extension of the concept. For example, the story, function machine, and recipe metaphors all encourage students to read the equals sign as a one-way sign connecting a process with an answer; the balance does not make sense when an equation contains negative quantities. Students need to know the limitations and the applications of each metaphor so that they can select whatever features are helpful as a support for abstract thinking.

REFERENCES

Fischer, R. "Didactics, Mathematics, and Communication." *For the Learning of Mathematics* 8 (1988): 20–30.

Johnson-Laird, P. N. *Mental Models.* Cambridge: Cambridge University Press, 1983.

Kieran, C. "Learning the Structure of Algebraic Expressions and Equations." In *Proceedings of the Twelfth Conference of the International Group for the Psychology of Mathematics Education,* pp. 433–40. Veszprem, Hungary, 1988.

Lakoff, G., and M. Johnson. *Metaphors We Live By.* Chicago: University of Chicago Press, 1980.

MacGregor, M., and K. Stacey. "Cognitive Models Underlying Students' Formulation of Simple Linear Equations." *Journal for Research in Mathematics Education* 24 (1993): 217–32.

———. "Learning to Formulate Equations for Problems." In *Proceedings of the Twentieth International Conference for the Psychology of Mathematics Education,* vol. 3, edited by L. Puig and A. Gutierrez, pp. 289–96. Valencia: PME, 1996.

Osgood, C. E. "Where Do Sentences Come From?" In *Semantics. An Interdisciplinary Reader in Philosophy, Lnguistics and Psychology,* edited by D. D. Steinberg and L. A Jakobovits, pp. 497–529. Cambridge: Cambridge University Press, 1971.

Pimm, D. "Metaphor and Analogy in Mathematics." *For the Learning of Mathematics* 1 (1981): 47–50.

Stassen, L. *Comparison and Universal Grammar.* Oxford: Blackwell, 1985.

17

Epistemological and Metacognitive Factors Involved in the Learning of Mathematics: The Case of Graphic Representations of Functions

Maria Kaldrimidou

University of Ioannina

Andreas Ikonomou

Technical and Vocational Teacher Training Institute
Athens, Greece

ATA and findings of many research reports (Artigue 1992; Bell and Janvier 1981; Clement 1989; Janvier 1987; Kaldrimidou and Ikonomou 1992; Monk 1992; Tall and Bakar 1992) point out the difficulties that students and teachers have in using, understanding, interpreting, and communicating through graphic means and representations.

This chapter deals mainly with the graphic representation of functions as a source of difficulties for many students and teachers of mathematics.

One could argue that students and teachers have difficulties with, and avoid, graphic representations because they have not learned how to use them. Classroom experience shows that this is not the reason. For example, efforts made in our research to modify the first-year calculus course to emphasize graphic representations proved to be insufficient in overcoming the students' difficulties. Moreover, the behavior of students during these courses suggests a certain reluctance to use graphic representations: "Not a figure again! Let's do some more calculating! Let's use formulas!" Moreover, research reports (Artigue 1992; Eisenberg 1992; Dreyfus 1991; Vinner 1989) stress that students avoid graphic representations in problems involving functions. Reviewing the relevant literature, Eisenberg reports that "these data show the reluctance students have to think visually" and "students have a strong tendency to think of functions algebraically rather than visually ... even if they are explicitly and forcefully pushed toward visual processing" (1992, pp. 164–65).

Cognitive reasons related to the specific characteristics of the language of graphs and epistemological reasons related to the beliefs about the nature of mathematics and the status of graphic representations in mathematics are the main explanations for students' difficulties with, and the avoidance of, graphic representations. The point of view expressed in this chapter is that there are also some conceptual reasons related to ideas about the function concept, the representations of this concept, and the procedures used in problems involving functions.

The first section of this chapter discusses the function concept, the different ways of representing functions, and their role in concept formation. The second section analyzes the graphic way of representing a function and its specific characteristics that distinguish it from other symbolic or schematic representations used in mathematics. The third section presents results from an empirical study that stress the link between conceptions of function and procedures used in tasks involving the production and interpretation of graphs. More precisely, this chapter shows that the procedures used in these tasks and the conceptions that students hold about functions are among the factors that could explain the students' difficulties and their reluctance to use graphic representations. Finally, the nature of these conceptions is discussed, along with the claim that they are a direct result of the communication in the classroom.

FUNCTION AND THE WAYS OF REPRESENTING FUNCTIONS

The concept of function holds a central position in science and in mathematics education. A function can be conceived of in various ways according to the place a function occupies in a certain context, that is, according to the problem in which it is involved, the ways of representing it, and the procedures (theorems, algorithms, tools) used with it. It must be pointed out that for the function concept, we have many different conceptions, depending on the context in which the concept is applied. For example, a function can be considered as a relation, as a transformation, as a mapping, or as an object. These conceptions are different from an epistemological point of view because the meaning of the notion is different in each situation. Thus, the representations and the procedures used in each situation are also different. Accordingly, we suggest that these procedures constitute the core of the notion of function in that they play an important role in the construction of this concept. The pointwise procedure, for instance, implies a local conception for the graph, which could explain the difficulties and the errors in problems involving an across-time interpretation of the geometrical features of the graphs (Monk 1988, 1992). By the same token, the operations that take place in the second member of the analytic expression of a given function (see MacGregor, chapter 16 in this volume) could imply an action-operation conception for the function (Breidenbach, Dubinsky, and Hawks 1992; Dubinsky and Harel 1992). The multiple character of this concept causes many problems in teaching and learning mathematics. A rich literature exists about the conceptions and misconceptions that students and teachers hold in relation to the notion of function, even when we limit ourselves to the case of a real function of a real variable (Even 1990; Monk 1988; Sierpinska 1992; Vinner 1983; see also more than 200 articles cited in Harel and Dubinsky [1992]).

One of the problems that the existing studies point out concerns the various ways of representing a function (algebraic or symbolic, numerical or tabular, graphical) and the difficulties that students have in establishing connections among them (Artigue 1992; Eisenberg 1992). Concerning the ways of representing a function, we must say that each way of expression highlights some aspect of a function and provides information regarding other aspects of this function that can be thus determined. But, each of the aforementioned expressions of a function cannot explicitly describe it in its entirety. Instead, they complement one another, and each one is a way of representing one or another aspect of the function. This becomes evident if one considers the terminology regarding functions. We talk about the formula of the function, its graphic representation, or the table-value of the function. Consequently, the means of representing the function cannot be deemed to be a symbol of the function, and it does not stand for the function.

As Sierpinska (1992) points out, some of the conditions of understanding functions are the following:

- "Discrimination between different means of representing functions and the functions themselves" (p. 53)
- "Synthesis of the different ways of giving functions, representing functions and speaking about functions" (p. 54)

Moreover, students very often identify functions with just one of the representations. They view functions in only one way and use only one mode of approach, usually the algebraic one, during their attempts to solve problems (Monk 1992; Vinner 1989).

Many researchers emphasize the role that the mastery of the graphic representation of a function and the flexibility in establishing relations among different ways of representing a function (especially between the algebraic and the visual) could play in the conceptualization of the notion of function (Sierpinska 1992; Artigue 1992; Eisenberg 1992).

GRAPHIC REPRESENTATION OF A FUNCTION

As already stated, the graphic representation of a function is just a way of representing the function and is not its symbol. However, a graph is a schematic representation. As such, it presents many difficulties arising from the specific characteristics of the language of a schematic representation, that is, difficulties of a cognitive nature.

As pointed out in the introduction, students often avoid proceeding graphically when they face problems involving functions. Beliefs and conceptions of an epistemological nature and of a metacognitive nature (about the difficulty or the efficacy of graphic procedures) could explain the students' attitude. These aspects are analyzed in the following sections.

The Cognitive Aspect

A graph is a schematic representation. As such, it presents many difficulties pertaining to decoding the information it contains. The source of these difficulties is the special way in which a graphic representation encodes information. The holistic character and the selective and economic way of data encoding are among these specific features (Kaldrimidou 1990). Many difficulties arise from this cognitive character of these representations: "Unfortunately, the ability to attend selectively and sequentially to parts and whole does not come easily for many students" (Yerushalmy and Chazan 1990, p. 201). In the same article, the authors identify three obstacles: the special character of a diagram, the existence of certain diagrams that become stereotypical (a fact that eventually causes problems whenever the students face a new diagram that does not conform to any of the stereotypes), and the inability to look at a diagram in different ways. Students must overcome these obstacles when examining and interpreting diagrams. These three obstacles impede the correct interpretation of a graphic representation.

The studies concerning the graphic representation of functions show that students do not usually find it difficult to identify and use correctly the graphic code. The pointwise treatment of a graph, that is, the identification of the coordinates of the points of a curve and the recognition of a discontinuity, for instance, do not pose any particular problems (Monk 1988). Reading and interpreting information are related to other features of a function, such as monotonicity and curvature, in contrast, raise considerable difficulties. The same holds for identifying the graphic representation of a piecewise function with a single function (Vinner 1989). In other words, it could be said that there are no decoding problems, only semantic ones.

Thus as far as the language of graphic representation is concerned, the following points can be noted:

1. The transition from the algebraic expression of a function to its graphic one, and vice versa, is not a simple translation (as Janvier [1987] points out), but a transposition. That is, a body of information is not converted to another symbolic system but is analyzed; this analysis yields new information, which in its turn is expressed in the other symbolic system. For example, the coefficients a, b, and c in the function y of x, given by an equation of the type $ay + bx + c = 0$, are real numbers, whereas the ratio $-b/a$ yields the slope of the straight line that the foregoing function represents.

2. An algebraic expression of a function is propositional in the sense that it carries information linearly by way of a sequence of sentences that can be read one after the other. The graphic representation of a function, however, is holistic. This means that the relations between the simple constituents of the graphic representation are simultaneously given and that its processing requires the analysis of the whole and the synthesis of the part.

3. Processing the graphic representation activates spatial-iconic

abilities, whereas the algebraic representation activates quantitative-relational abilities (Demetriou and Efklides 1994). To the extent, therefore, that teaching in school does not create the educational environment necessary for the development of the spatial-iconic ability, it is only natural that the students' comprehension and use of the graphic representation leave much to be desired. The complicated nature of the graphic representation and its spatial-iconic character make its processing depend on the cognitive profile of the subject (field dependence or independence, for instance).

4. The graphic representation implicitly contains more information than the information needed to solve a given problem. Thus, processing a graphic representation requires selecting those data that are necessary for this solution. In consequence, the possible differences between interpretations of the same schema are greater than the same differences regarding information represented in other symbolic systems.

The Epistemological and Metacognitive Aspects

This discussion attempts to shed some light on the misunderstandings that are observed when the subject matter of the graphic representation of a function is at issue, but it cannot explain the avoidance and reluctance on the students' part toward graphic representations and even the anxiety that students feel when facing tasks regarding graphic representations, as Ruthven (1990) remarks.

Dreyfus (1991) and Eisenberg (1992) stress that visual (and thus graphic) reasoning is considered to be of a low epistemological status both in mathematics and in mathematics education. In other words, "visual is not mathematical" (Eisenberg 1992, p. 167).

Artigue (1992) reports that proofs were more easily understood in an algebraic than a graphical setting and that for graphic proofs, several difficulties arise concerning the efficiency and the rigor of the justifications.

In questions that ask the student to choose between two kinds of answers (Kaldrimidou 1995)—algebraic or schematic—the majority of students and teachers (from 68.3 percent to 95 percent for the students and from 61.5 percent to 69.2 percent for the teachers) prefer the algebraic ones. (These questions were presented to the same group of people employed in the empirical study reported in this chapter.) They did so because the algebraic procedure is more mathematical, more general, more complete and exact; is more comprehensible and easier to treat; has less risk of an error; is safer.

Thus, the reported avoidance of visualization is related, in an implicit way, to epistemological conceptions (concerning the nature of mathematics and of mathematical procedures) and metacognitive conceptions (concerning the efficiency and the ability to use procedures).

THE QUESTIONNAIRE

In summary, the previous analysis indicates that the difficulties of students and teachers are caused by the characteristics of the graph as a schematical-geometrical representation and as a diagram and by some more general beliefs and conceptions concerning the nature of mathematics and the nature of procedures used in mathematics. But we can argue that these reasons are valid for any schematical representation because they do not take into account any specific characteristic of the graph on a conceptual level, that is, as a representation of a function.

As a representation of a function, the graph maps the functional relation into a geometrical setting. Thus, for the graph to function as a representation, the geometrical features must be read appropriately. To see only the shape of the curve is a restriction, an inadequate way to read a graph. We must see that points on the curve are symbols that represent the value of the independent variable, the value of the dependent variable, and the relationship between them. So we could say that a good mastery of the graphic representation as a means of representing functions depends on understanding a function.

This is the main idea that the following empirical study attempts to assess. A questionnaire was given to students and teachers of mathematics (sixty university mathematics students and twenty-six high school mathematics teachers). It included three groups of items:

Group 1: This group included three questions. These were questions outside the usual graphic context, formulated in verbal form with no reference to graphic representation, which could be elaborated graphically or not. The intention was to find out if the use of a graphic representation requires a particular context for recognizing the corresponding field (a particular field recognition context).

Group 2: This group included five items that asked explicitly for a graphic representation of a function, the construction of a geometrical figure (picture), or both. They were typical school mathematics problems.

Group 3: This group included four items that asked for an interpretation of a graphic representation.

An indication of the reliability of the questionnaire was that teachers proved to be more expert than students in using complex tools (i.e., integration), and so they made fewer errors. But the analysis focuses on the ways that teachers and students conceive of the graphic representation, the procedures they use, and their attitude toward it as a mathematical tool.

The answers of both students and teachers point to a general negative attitude toward the use of graphs, whereas their reaction is positive vis-à-vis the algebraic procedures.

For example, consider the question in figure 17.1.

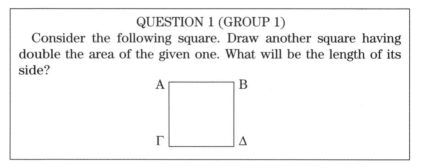

Fig. 17.1

Most of the answers had the following characteristics:

- Reversal of the order of the questions
- Typical demonstration, often exaggerated, for the second question
- Design of a square, not related to the given one

The responses in figures 17.2, 17.3, and 17.4 illustrate the foregoing remarks. Answers of the type in figure 17.4 were rare. This item is not an example for showing the propensity for one type of solution (algebraic) over the other (schematic). It is, however, an example of the fact that students and teachers of mathematics react differently to questions involving algebraic procedures than they do to questions involving schematical procedures. Students and teachers of mathematics do not have to choose between two types of solution. They have to answer two questions: first, they must find the length of the square with double

EXAMPLE OF AN ANSWER TO QUESTION 1: CASE 1

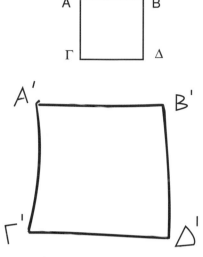

Area of a square = a^2, where a is its side, so if a_1 is the side of the primitive square AB$\Delta\Gamma$ and E_1 is its area, then $E_2 = 2E_1 = 2a_1^2$, $E_2 =$ the area of the second square, so

$$a_2 = \sqrt{E_2} = \sqrt{2(a_1)^?} = \sqrt{2}a_1.$$

Fig. 17.2

EXAMPLE OF AN ANSWER TO QUESTION 1: CASE 2

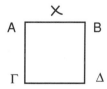

$$E = \chi^2$$
$$2E = 2\chi^2$$
$$2E = \left(\sqrt{2}\chi\right)^2$$

Fig. 17.3

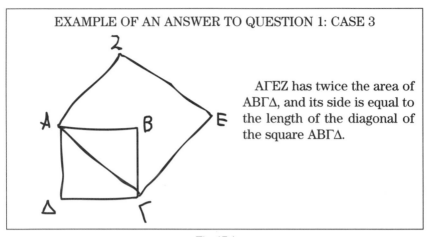

Fig. 17.4

the area, and, second, they must construct this square geometrically. Their answers show that the majority of the students and teachers paid little attention to the question asking for a geometrical construction (i.e., a drawing), whereas they demonstrated very carefully the question requiring an algebraic approach. In the given examples (case 1 and case 2), there is no indication of the construction of the square A′B′Δ′ Γ′ with double the area of the given one. Moreover, careful examination shows that the constructed squares bear little resemblance to a square with double the area of the given one. The area of the square A′B′Δ′ Γ′ in case 1 is greater than three times the area of the given square; in case 2, the area of the constructed square is approximately 2.5 times the area of the given square. Moreover, concerning the other items of the first group with verbal contexts, no student and no teacher but one used a schematic procedure to solve the problem, although the whole questionnaire focuses on using graphic representations.

THE NATURE OF THE GRAPHIC REPRESENTATION

Two questions, one for group 3 (fig. 17.5a) and one for group 2, are next presented schematically. Typical answers are given to each question to illustrate the students' and teachers' conceptions.

Group 3's answers (figs. 17.5b, c, d) provided information regarding how teachers and students conceive of the graph as a way of representing

QUESTION 2 (GROUP 3)

Report all the information you can get for the function whose graphic representation is given in the following figure:

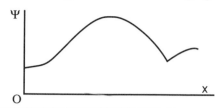

Fig. 17.5a

EXAMPLE OF AN ANSWER TO QUESTION 2: CASE 1

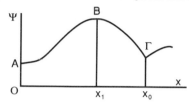

The function is continuous and has a derivative at all points of its domain not, except the point χ_0, which corresponds to point Γ. The function has a minimum at point A and a maximum at point χ_1 and also has local minimum at point Γ. Finally, I can draw conclusions for the concavity of the functions.

Fig. 17.5b

EXAMPLE OF AN ANSWER TO QUESTION 2: CASE 2

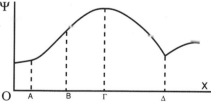

It is continuous, defined for $\chi > 0$, it is positive ($\Psi(\chi) > 0$).

Point B is a point of inflection. Point Γ is a maximum (as far it is shown here to be global). Point Δ is a minimum. For the monotone property, I have—

O —> Γ: Rising (O —> B: upward, B —> Γ: downward),

Γ —> Δ: Falling (downward),

Δ —> ∞: Rising (downward).

Fig. 17.5c

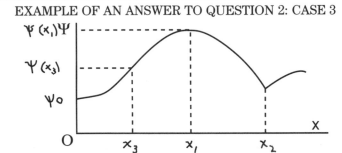

The domain of the function is the set $[0, +\infty)$, and its set of values is the interval $[\Psi_0, \Psi_1]$, $(\Psi_1 = \Psi(\chi_1))$.

The function has a local maximum at $(\chi_1, \Psi(\chi_1))$, where for this value the first derivative is equal to zero and the second derivative is negative.

The point $(\chi_3, \Psi(\chi_3))$ is a point of inflection, that is to say, the second derivative is equal to zero.

Fig. 17.5d

a function and, more generally, gave information regarding their conception of the nature of a function.

Three different ways of conceiving of the graph were noted in these three cases that correspond to three different conceptions of functions:

Conception 1: The curve is considered independently of the system of axes. Reading is done from left to right, and the variations are described as if the student had actually followed the trace of the curve in the plane (i.e., the function is ascending to point A, descending to point B, then ascending again). Here, the designation of the characteristic points is made on the curve with a single letter, and what is perceived is the shape of the curve, the axes practically being ignored.

Conception 2: The curve is considered only in relation to the X-axis: The curve is separated in parts, which are defined by the characteristic points set by the value of the independent variable on the X-axis. Every zone is read without the others being taken into account. Usually this way of understanding is accompanied by a presentation of the general characteristics of a function (domain of the function, range of values, continuity, monotone property, concavity). It could be said that in this case, the students and the teachers are influenced by the standard procedure of studying a function.

Conception 3: The curve is considered in relation to the two axes. The points have two coordinates, and the way in which the information is given reflects a mature understanding of the function.

All three conceptions are equivalently distributed in the minds of the students, although the teachers present an important accumulation around the second conception (73.1 percent). This difference would appear to be a good example of how the content of the mathematical target in school affects the epistemological conceptions of teachers.

THE PROCEDURES

The question for group 2 is shown in figure 17.6a

QUESTION 3 (GROUP 2)

Sketch the graphic representation of the function defined by $\Psi(\chi) = \chi^2 - 2\chi + 1$.

Fig. 17.6a

It can be noted that students and teachers have used three procedures to study and represent a function graphically.

1. The point-by-point procedure (fig. 17.6b): Students (teachers) have found some characteristic points (intersection points, coordinates of minima and maxima).

2. The step-by-step procedure (fig. 17.6c): Students (teachers) have used the typical procedure taught in school (monotone property, concavity, table of variation).

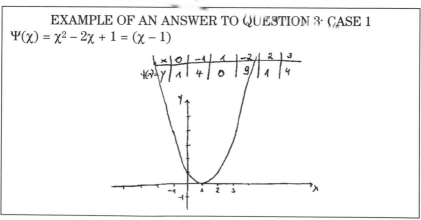

EXAMPLE OF AN ANSWER TO QUESTION 3 CASE 1
$\Psi(\chi) = \chi^2 - 2\chi + 1 = (\chi - 1)$

Fig. 17.6b

EXAMPLE OF AN ANSWER TO QUESTION 3: CASE 2

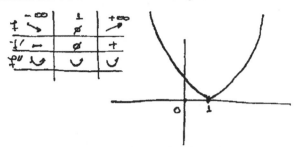

We shall find the roots of $\Psi(\chi) = \chi^2 - 2\chi + 1$

$$\frac{2 \pm \sqrt{4-4}}{2} = \frac{2}{2} = 1, \text{ double root.}$$

The derivative. $\Psi(\chi) = 2\chi - 2$ with root $2\chi - 2 = 0$, $\chi = 1$, so the point $\chi = 1$ is maximum or minimum of $\Psi(\chi)$.

The second derivative $\Psi''(\chi) = 2 > 0$, so the direction of concavity (of $\Psi(\chi)$) is upward.

Fig. 17.6c

3. The holistic procedure (fig. 17.6d): Students (teachers) have used global characteristics, such as the fact that the function is represented by a parabola, and they used the method of changing the system of coordinates.

EXAMPLE OF AN ANSWER TO QUESTION 3: CASE 3.

$\Psi(\chi) = \chi^2 - 2\chi + 1 = (\chi - 1)^2$

$\omega = 1$

$\Psi(\omega) = \omega^2$

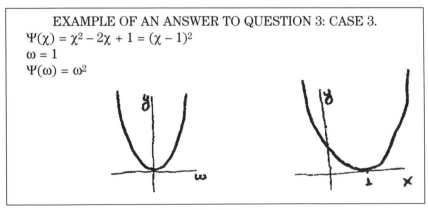

Fig. 17.6d

This question shows a similar distribution between students and teachers regarding the way in which they have used the three procedures. Teachers present a consistency in the choice of the procedures they have used, which was confirmed by means of a factor analysis. Nevertheless, the preference of the two groups is centered on the point-by-point procedure (45.6 percent for the students and 47.1 percent for the teacher; the percents refer to only the persons who answered the questions).

How could this preference be explained? The procedure of finding values occupies a very important place in school lessons. Indeed, finding the characteristic points and values is an important matter in many mathematical fields (equations and systems solutions, roots of polynomials, limits of sequences, intersection points of geometric figures) and, of course, in school mathematics. On the one hand, finding the characteristic values is often the final target of the whole activity.

On the other hand, finding the characteristic values is a relatively simple algebraic computational activity. As such, it allows a facile error control. Sierpinska (1992, p. 46) considers the preference for this kind of procedure as an obstacle to the comprehension of the concept of function.

Therefore, the importance of this procedure and its easy method of error control have granted it a leading position among the epistemological and metacognitive conceptions of students and teachers.

What are the repercussions of these conceptions on the conceptualization of the notion of function? As already mentioned, the pointwise conception and the identification of a function with its analytic expression do not impose only a restricted conception of a function, but they could also be viewed as "mécanismes producteurs d'obstacles" (Artigue 1991, p. 261), if we consider that they have become a dominant way of conceiving of function and that they have been deemed, by the students, to be the exemplar of mathematics par excellence.

FINAL REMARKS

Summarizing, we note that students and teachers do not pay much attention to pictures and graphs; they use mainly algebraic procedures, and only a minority understand the graph as a graphic representation of a function in the full sense of the term. The following points can be made:

1. The difficulties observed in the use and interpretation of graphs are due not so much to factors that are related to the language of the graphs (code), but to the significance that this code has in the context

of school lessons (for the semantic of a text, see also the contribution of Navarra, chapter 19 in this volume). That is, these difficulties are related to—

(*a*) general epistemological characteristics (the mathematical value of the graphic representation and of the algebraic procedure);

(*b*) specific epistemological characteristics (the geometrical and point-by-point conception of a curve versus the functional relation it represents);

(*c*) metacognitive characteristics (the simplicity and effectiveness of various procedures).

These conceptions are created in the classroom as a result of teaching. More specifically, they are the result of the material taught in class and of the problems solved by the students. They also result from the way in which the efficiency of students is evaluated in the relevant problems. Thus, graphic representations constitute, for the majority of teachers, one-tenth of the total credit attributed to a given problem.

2. The whole context in which functions are approached and taught favors, indirectly, the algebraic procedures, which, considering the special weight of a graph as a representation of function, is not the most appropriate way of presenting the matter. This algebraic context causes students to develop a conception of function according to which the graphic representation is an accessory and not one of its main constituents.

3. The subject of mathematics, as taught in school, gives rise to epistemological and metacognitive considerations that favor the algebraic rather than the graphic representation of problems. This means the creation of an implicit context of communication that reinforces the algebraic rather than the graphic representation of a function. This process produces a vicious circle, in the sense that the solution of a problem by means of an algebraic representation results in a reinforcement of this representation with respect to its effectiveness (metacognitive characteristic). The continuous repetition, and its reinforcement in school, strengthens this kind of representation and makes it the prominent and most appropriate procedure in mathematics (epistemological characteristic).

These considerations make it clear that not only the pedagogics, but also the material taught in class, influences and shapes the context of communication among students, thus among students and teachers, as well as the relationship between students and the material itself. How this material is taught eventually establishes the epistemological, cognitive, and metacognitive profiles that work successfully in only a very limited context. These profiles function as filters that create communication problems, which, in their turn, form conceptions that differ from the mathematical concept.

4. The foregoing analysis shows how mathematical conceptions of students can be manipulated and reconstructed during the teaching process.

REFERENCES

Artigue, M. "Epistémologie et didactique." *Recherches en didactique des mathématiques* 10 (1991): 241–85.

———. "Functions from an Algebraic and Graphic Point of View: Cognitive Difficulties and Teaching Practices." In *The Concept of Function: Aspects of Epistemology and Pedagogy*, edited by G. Harel and E. Dubinsky, pp. 109–32. Washington, D.C.: Mathematical Association of America, 1992.

Bell, A., and C. Janvier. "The Interpretation of Graphs Representing Situations." *For the Learning of Mathematics* 2 (1981): 34–42.

Breidenbach, D., E. Dubinsky, and J. Hawks. "Development of the Process Conception of Function." *Educational Studies in Mathematics* 23 (1992): 247–85.

Clement, J. "The Concept of Variation and Misconceptions in Cartesian Graphing." *Focus on Learning Problems in Mathematics* 11 (1989): 77–87.

Demetriou, A., and A. Efklides. "Structure, Development, and Dynamics of Mind: A Meta-Piagetian Theory." In *Intelligence, Mind, and Reasoning: Structure and Development*, edited by A. Demetriou and A. Efklides, pp. 75–110. Elsevier Science D.V., 1994.

Dreyfus, T. "On the Status of Visual Reasoning in Mathematics and Mathematics Education." In *Proceedings of the Fifteenth Conference of the International Group for the Psychology of Mathematics Education*, pp. 33–48. Assisi, Italy, 1991.

Dubinsky, E., and G. Harel. "The Nature of the Process Conception of Function." In *The Concept of Function: Aspects of Epistemology and Pedagogy*, edited by G. Harel and E. Dubinsky, pp. 85–106. Washington, D.C.: Mathematical Association of America, 1992.

Eisenberg, T. "On the Development of a Sense for Functions." In *The Concept of Function: Aspects of Epistemology and Pedagogy*, edited by G. Harel and E. Dubinsky, pp. 153–74. Washington, D.C.: Mathematical Association of America, 1992.

Even, R. "Subject Matter Knowledge for Teaching the Case of Functions." *Educational Studies in Mathematics* 21 (1990): 521–44.

Harel, G., and E. Dubinsky, eds. *The Concept of Function: Aspects of Epistemology and Pedagogy.* Washington, D.C.: Mathematical Association of America, 1992.

Janvier, C. "Translation Processes in Mathematics Education." In *Problems of Representations in Mathematics Learning and Problem Solving,* edited by C. Janvier, pp. 27–31. Hillsdale, N.J.: Lawrence Erlbaum Associates, 1987.

Kaldrimidou, M. "Images mentales et activité mathématique." *Cahiers de didactique des mathématiques* 4 (1990): 21–47.

———. "Lo status della visualizzazione presso gli studenti e gli insenanti di matematica." *La matematica e la sua didattica* 2 (1995): 182–94.

Kaldrimidou, M., and A. Ikonomou. "Aptitude des bacheliers à manipuler des représentations graphiques." *Cahiers de didactique des mathématiques* 10–11 (1992): 159–81.

Monk, S. "Students' Understanding of Functions in Calculus Courses." *Humanistic Mathematics Network Newsletter* 2.(1988).

———. "Students' Understanding of a Function Given by a Physical Model." In *The Concept of Function: Aspects of Epistemology and Pedagogy,* edited by G. Harel and E. Dubinsky, pp. 175–93. Washington, D.C.: Mathematical Association of America, 1992.

Ruthven, K. "The Influence of Graphic Calculator Use on Translation from Graphic to Symbolic Forms." *Educational Studies in Mathematics* 21 (1990): 431–50.

Sierpinska, A. "On Understanding the Notion of Function." In *The Concept of Function: Aspects of Epistemology and Pedagogy,* edited by G. Harel and E. Dubinsky, pp. 25–58. Washington, D.C.: Mathematical Association of America, 1992.

Tall, D., and M. Bakar. "Students' Mental Prototypes for Functions and Graphs." *International Journal of Math, Education, Science and Technology* 23 (1992): 39–50.

Vinner, S. "Concept Definition, Concept Image and the Notion of Function." *International Journal of Math, Education, Science and Technology* 14 (1983): 293–305.

———. "The Avoidance of Visual Considerations in Calculus Students." *Focus on Learning Problems in Mathematics* 11 (1989): 149–56.

Yerushalmy, M., and D. Chazan. "Overcoming Visual Obstacles with the Aid of the Supposer." *Educational Studies in Mathematics* 21 (1990): 199–219.

18

Making Mathematics
Accessible

Megan Clark
Victoria University
Wellington, New Zealand

THE language used in the exposition, assessment, and problems of mathematics can make mathematics more, or less, accessible to the learner. In this chapter, the success of students in solving problems set in different contexts and their choice of contexts, when one is available, are examined in two large sample research studies: one at the secondary school level and one at the first-year university level. The outcomes for different groups of students, in particular males and females, are compared. The language that seems likely to make mathematics more accessible to young women is that relating to the personal experience of the learner. The familiarity and the relevance of the context of a problem both seem to have an effect on whether the problem is attempted and how successful that attempt is. The results of the studies reaffirm the importance of the need for teachers to exercise sensitivity in their choice of the appropriate language to use with their specific students.

CONTEXT AND PERFORMANCE IN
SECONDARY SCHOOL MATHEMATICS

In New Zealand in the last decade, there have been large increases in the proportion of the age cohort going on to senior secondary school, as shown in table 18.1. The first year of secondary school is known as Form 3 in New Zealand, and the last is Form 7.

Table 18.1
Secondary School Retention Rates from Form 3
(Grade 9) through Form 7 (Grade 13)

Year	Percent
1980	14.6
1991	39.9

The percent of young women taking mathematics at sixth form (grade 12) has increased from 55 percent to 76 percent from 1974 to 1991 (New Zealand Ministry of Education 1992). This increase is in part because of social pressure to continue with mathematics and in part because of recent efforts by schools to make mathematics and statistics as accessible and as relevant as possible to as many students as possible. The basic concerns of mathematics teachers at every level are the same: how to find the ways to best teach these increased numbers of students, especially the groups not previously well served by the mathematics curriculum, in particular, female, Pacific Island, and Maori students.

When Kath Hart, noted British mathematics education researcher, visited New Zealand in 1989 she said in an address to Wellington College School of Education, "We underestimate the effect of the way we write questions," illustrating her statement by the following example:

> Brown, Smith, and Jones come in to work 2, 4, and 6 days a week, respectively. The total coffee bill for the week is £6. How much should each pay?

In versions 2 and 3, the following changes were made: "the total electricity bill" and "the cost of ham rolls." British children found the first version the most difficult, and the third, the easiest.

The language and context of problems, activities, and questions in mathematics classes seem to be crucial to our pupils' success. Of particular interest in this regard are the New Zealand Form 3 (grade 9) results of certain questions from the 1981 Second International Mathematics Study (SIMS) (Wily 1986) as shown in table 18.2.

Other questions showed a similar pattern. A question about a male painter had boys ahead by 7 percent, and one on the number of students in a school had girls ahead by 3 percent.

Part of the 1981 SIMS study at the grade 9 level was repeated by the Equity in Mathematics Education group in New Zealand in 1989 with a sample of 800 pupils. The average total score for female pupils on a test of forty questions improved significantly from 1981 to bring them level with the boys (Forbes et al. 1989). However, the area of word problems

Table 18.2

Selected Results from the 1981 Second International Mathematics Study

	Percent Correct	
	Girls	Boys
A model *boat* is built to scale so that it is 1/10 as long as the original boat. If the width of the original boat is 4 metres, the width of the model should be (0.1m, *0.4m,* 1m, 40m).	56	66
Cloth is sold by the square metre. If 6 square metres of cloth cost $4.80, the cost of 16 square metres will be (*$12.80,* $14.40, $28.80, $52.80, $128.00).	51	48
How many pieces of *pipe* each 20 metres long would be required to construct a pipeline 1 kilometre in length? (5, *50,* 500, 5 000, 50 000)	38	50
A 15-centimetre piece is cut from a *ribbon* 1 metre long. What is the length of the remaining piece? (*85 cm,* 115 cm, 985 cm, 1015 cm, 9985 cm)	78	76

was not an area of improvement for the girls. Some of the disappointing results are demonstrated in table 18.3.

Table 18.3

Male Percent Success over Females in Selected Questions from the 1981 Second International Mathematics Study (New Zealand Sample)

	Boys ahead by	
Question content	1981	1989
Q. 12 male painter	7	8
Q. 14 model boat	10	11
Q. 24 cloth buying	–3	1
Q. 33 number of pupils	–3	1

There is a suggestion here that boys have learned to cope with questions about cloth in the intervening years, whereas the girls have not yet been able or been willing to cope with questions about model boats.

New Zealand girls seem to be doing better at problems that, as Howson and Mellin-Olsen (1986) describe, have meaning and significance for them. This implies that we should present problems that make female students feel more at home in their mathematics lesson. But male students are doing just as well in problems from the female domain as they are with those from traditionally masculine settings. Surely it is desirable that female students should be as domain-free as the males and not be restricted to problems of cookery and sewing. Like all real educational problems, this issue is complicated. Probably, in the first instance, it is wise to give girls more familiar material, more home-based examples, because until they are convinced that mathematics is just as important a part of their world as it is of their brothers', they may not make much more headway. They are in the same position as the boys were in 1981, when they did much better on the pipe problem than on the ribbon one. The need now is to convince the girls that this material is accessible and important to them.

FIRST-YEAR UNIVERSITY STATISTICS EXAMINATIONS

Similar phenomena have been observed in New Zealand at the university level. Victoria University has two different first-year statistics courses:

- STAT 131 Data and Probability, the course recommended for students majoring in mathematics, physics, chemistry, computer science, and engineering

- STAT 193 Statistics for the Natural and Social Sciences, suggested for those majoring in biological sciences, social sciences, commerce, and medicine.

Results for the last few years are given in table 18.4.

It would seem easier, in general, for females to succeed in the STAT 193 course than in the STAT 131 course, since they consistently earn a larger percent of the A grades than their percent presence in the class, whereas the converse is sometimes true for STAT 131.

Comparing the content and results for examinations in these courses is instructive. In the 1989 STAT 131 final examination, the males performed better on five of the six questions. Each of these six questions had two parts. Males performed better on the parts that were abstract (there were six of these, whereas there were four that involved people or animals and two that dealt with machinery; one of the people questions also involved cars). This result suggests that an imbalance in question content may be affecting results. In 1990, the whole STAT 131 examination was somewhat abstract. No significant differences occurred in the total examination score by gender, although males did perform better. Taken question by question, only question 5 stands out: a significantly higher percentage of males than females opted to try this

Table 18.4

Participation and Success for Female Students in First-Year Statistics Courses

Year	Percent female		Female percent of A grades	
	STAT 131	STAT 193	STAT 131	STAT 193
1988	30	36	27	48
1989	21	41	25	59
1990	37	52	37	57
1991	28	54	38	60
1992	29	64	20	73
1993	25	59	31	74

Notes:

1. A deliberate effort was made to advise and encourage females into STAT 131 in 1990. It is of concern that the proportion of females in STAT 131 has only with particular effort approached that of females in general mathematics courses, such as algebra, geometry, and calculus, for which the proportion is usually 30 percent to 35 percent.

? A small group of students takes both courses. For such students, their grades in both courses are remarkably close, usually within 3 percent, so there is no reason to suspect that one course is markedly easier than the other.

question, and they did significantly better on it. This question was about the costs of building ferries, expected profits, standard deviation, and expected loss (see fig. 18.1).

In the STAT 193 examination for that year, male and female overall scores were not different, but some interesting points emerged. Gender was not divisive. Males and females opted to try the same questions.

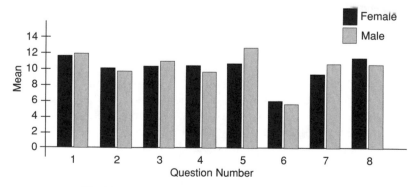

Fig. 18.1. 1990 STAT 131 grades by question and gender

Questions 1, 6, and 7 were the most popular. These questions were about histograms of numbers of children in a family, an analysis of variance involving insurance sales, and a contingency table of vaccination levels versus ethnicity. Females also liked (that is, chosen by more than 80 percent) a question on regression that involved bacteria mutations. Only one question had a significant difference in scores—number 7—which was about vaccination levels and ethnicity. Females did better on this question.

Some very abstract questions were included on this examination—questions 3, 4, and 5—about finite populations and the central limit theorem, the derivation of variance and mean, and the derivation of power. These questions were not popular choices.

The students were most successful in the questions that were chosen by the highest percentage of the class. Either they are good at what they like or they know what they are good at. It was noticeable that not just female students responded well to the people questions; males did also, as shown in figure 18.2.

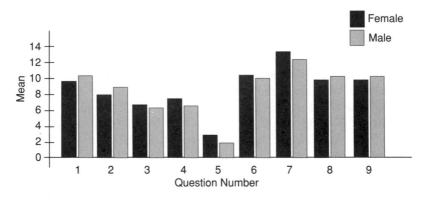

Fig. 18.2. 1990 STAT 193 grades by question

In 1991 the STAT 131 examination was less abstract, and females were slightly ahead in overall examination scores. Question by question, only question 2 stands out. Females did significantly better on this question, which was about a multiple-choice test and the probability of various grades and a second part about the mean area of some fencing—a question following a very familiar calculus path. There was a slight female preference for question 6, which was a very standard piece of theory.

The STAT 193 examination for 1991 showed a pattern similar to that of the previous year. In general, males and females chose the same questions. Questions 6, 7, and 9 were the most popular: an analysis of variance question on IQ, a contingency table analysis on spouse abuse versus family type, and a regression question relating cave art to animal remains. Questions 3 and 4 were more technical and abstract than the rest of the examination; they were the least popular and the least successfully done. Four questions had a significant difference in scores. Questions 5, 6, 7, and 9 were questions in which female students did better than males.

Questions 5 and 7 were chosen by a significantly higher percentage of females than males, and questions 1 and 2 were chosen by a higher percentage of males, who then did better than females in question 2, but not significantly (see fig. 18.3). Overall, females did significantly better in this examination than males did.

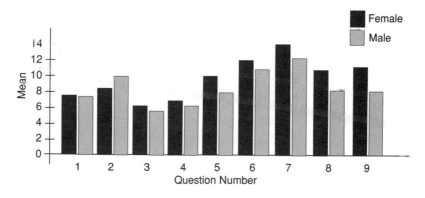

Fig. 18.3. 1991 STAT 193 question means according to gender

THE STUDENTS

In general, male students in STAT 193 are not males doing standard mathematics courses. The majority are doing no other mathematics and may not even have done mathematics as far as the seventh form (grade 13). These males are not studying in technical or mathematical areas, and it seems clear that the problems that appeal to these males—who have mostly rejected, or not been successful in, mainstream mathematics—are, by and large, the problems that appeal to the females in the class.

The numbers of Maori (indigenous New Zealanders) in these classes are too small (4 and 15 in STAT 131 and STAT 193, respectively) to draw any statistically significant conclusions, but some indications exist. In 1991 the STAT 193 class contained a small group (20) of very mathematically able, highly skilled premedical students. With the exception of one Cambodian, these were all Pakeha (European) New Zealanders. No Maori students were in this group of medical students; thus no Maori were among the highest scorers on any question and the Maori mean and median scores were lower than those of the Europeans.

Nevertheless, there was one question that a greater proportion of Maori students chose to answer than did the European students and on which some Maori students did very well. This was the regression question relating prehistoric cave art in France to the incidence of mammoth bones. Question 2, on beetles, was one on which average Maori performance was equal to that of the Europeans in the class.

It seems reasonable that certain questions are significant and meaningful to various ethnic groups and enable them to demonstrate their knowledge. It is this enabling power of the language of questions that is of primary importance.

OTHER ASPECTS OF THE COURSES

Success in the examination, however, is not entirely because of the content of the examination. The courses themselves have major differences. One is the context of the content. In these two courses, all assignment problems for two years were categorized into three groups: (1) abstract and traditional problems (urns, dice, $P(A \cup B)$, cards, coins, random numbers, and so on); (2) problems about people and animals (blood pressure, bacteria, and so on); and (3) a smaller category of nontraditional and nonabstract problems whose content could be seen as belonging to that sphere of activities traditionally thought to be appropriate to males (Morse code, football matches, stock market). Results are in table 18.5.

This difference in content suggests some reasons why female students might prefer STAT 193, since much research shows that "the most influential factor in girls' attitudes to study is whether the subject is perceived as a 'male' or 'female' subject" (The Royal Society 1986, p. 20).

Another suggestive difference is in the staffing of these courses. New Zealand universities do not have many female or Maori mathematics staff. Therefore, our European male students get many role models and signals that statistics is something that they can reasonably expect to be good at, whereas our female students do not. STAT 131 has three male lecturers, whereas STAT 193 has two male lecturers and one

Table 18.5
Numbers of Different Categories of Problem
in First-Year Statistics Courses

| | Number of Problems | | | |
| | STAT 131 | | STAT 193 | |
	Year 1	Year 2	Year 1	Year 2
Abstract or traditional problems	70	57	50	9
Nontraditional and nonabstract problems from the boys' world	10	21	7	5
People and animal problems	36	30	64	55

female lecturer. It is worth noting that on the 1990 examination, questions 3, 4, and 5 were on material lectured by a male staff member and questions 6, 7, and 8 were on material given by the only female lecturer involved in the course. In 1991, questions 3 and 4 were on material given by a male, and questions 5, 6, 7, 8, and 9 were on material given by the female staff member. Perhaps an effort should be made to offset this imbalance in the gender of staff by taking particular care with assignment questions.

CHOICE OF CONTEXTS

In 1991, further work was done with the STAT 193: Statistics for Natural and Social Sciences group. In one of their assignments they had the following question:

Do:
Either
A psychologist wishing to compare three methods for reducing hostility levels selected eleven high scoring (and roughly equal) students on a standardized test of hostility levels. Five students were treated by method A, three by method B, and three by method C. After a course of treatment, each student was tested again. The table gives the score, X, on the hostility level test after treatment.

Treatment	X	T_i	n_i	mean $\overline{X}_i$
A	73, 83, 76, 68, 80			
B	54, 74, 71			
C	79, 95, 87			
		$T = 840$	$n = 11$	

Complete the table. (Students are expected to fill in totals, means, and numbers of subjects in each treatment as a preliminary to an analysis of variance.)

Or

In a comparison of the strengths of concrete produced by three experimental mixes, five specimens were prepared from each type of mix. Each of the fifteen specimens was subject to increasing compressive loads until breakdown. Four specimens were tested incorrectly and their results had to be ditched. The table gives the compressive loads, X, in hundredweight a square inch, attained at breakdown. (Followed by table and question, as above.)

Very few students appeared to notice that these were the same question. Half the class had the question printed in this way, and the other half received it with the content of the Either and Or parts interchanged.

It is clear from table 18.6 that by and large, people answer the first option available. Of the 217 students, 166 took the first option on the paper, but if they were women students, they took this option more strongly when the first option concerned the psychologist. Or, if the first was about concrete, women were more likely to take the second option. In this case, women split about 50:50 between the options, whereas men stayed 2:1 with the problem about concrete. Overall, 73 percent of the females and 53 percent of the males took the psychologist option.

Table 18.6
Student Choices When Presented with Alternative Questions

	Option Chosen			
	Concrete		Psychologist	
	First	Second	First	Second
Male	39	5	32	17
Female	31	3	64	26

This result supported the idea that traditional methods of mathematics teaching draw very heavily on example material that is most effective for a particular subset of the population, usually males, with a very technical or abstract bias. At the present time, many more numerate people are needed in the workforce, and mathematical competence needs to be much more widespread. It is necessary for males without an engineering aptitude as well as for female, Maori, and Pacific Island students.

LANGUAGE AND LEARNING MATHEMATICS

It has been noted by Goodman (1986) that language learning in schools can be made easier or harder, shown by the following:

It is easy when—	*It is hard when—*
it is real and natural,	it is artificial,
it is whole,	it is broken into bits and pieces,
it is sensible,	it is nonsense,
it is interesting,	it is dull and uninteresting,
it is relevant,	it is irrelevant to the learner,
it belongs to the learner,	it belongs to someone else,
it is part of a real event,	it is out of context,
it has a social utility,	it has no social value,
it has a purpose for the learner,	it has no discernible purpose,
the learner chooses to use it,	it is imposed by someone else,
it is accessible to the learner,	it is inaccessible to the learner,
the learner has power to use it.	the learner is powerless.

This applies to the learning of language, but it is likely that it is just as applicable to mathematics. Unfortunately, many mathematics activities, such as worksheets, are more typical of the right-hand column than the left, which suggests why the same girls who deal daily with such knitting-pattern instructions as K1P2S1PSSO cannot or will not deal with something as comparatively straightforward as x/y.

Young children can only think about and comprehend mathematics as it relates to specific contexts (Reeves 1990). When it comes to pipes, cloth, ribbons, and x/y, it is easier for them to establish their hold on this material if they are considering it in a context that works for them. Then later, manipulating abstract entities like x/y might make real sense rather than be learned as a rote skill that does not work for them outside the very narrow confines in which they have previously been exposed to it.

Language is all important. The language of examples and problems can make those problems more or less accessible to the student. We can ensure that the language we use is not full of premature or unnecessary abstraction. We should be encouraging pupils to use language in their mathematics activities so that they get to feel some ownership over that activity.

Furthermore, the language used in assessment is vitally important. There is not much point in taking great care over the language used in instruction in the classroom if the language of assessment remains unaltered. Students should be able to demonstrate their mathematics skills in the language in which they function most comfortably. In particular, they should be able to be assessed in their own first language if that is their choice.

Careful listening to the language of our students can sometimes reveal a great deal about how a student is seeing a problem. This question is from a mature student about the decimal number, 0.35: "How do you turn it into a fraction?" Notice the language of magic—"turn it into" —which is exactly how this student viewed the mysterious manipulations in which the rest of the group were involved. Teachers also tend to slip into this language of magic, thus sending all the wrong messages to many of the pupils. Mathematics thus may appear to some as an arcane discipline for privileged initiates only.

Most students lack opportunities to express mathematical ideas in speech or writing (EQUALS 1989). It is exciting to see students keeping individual mathematics journals, being involved in group discussions of problems, and communicating what they know to others. All these activities, although useful in themselves, have the added benefit of undermining the inarticulate stereotype of the person who is good at mathematics, a picture that is unappealing to much of the population both at school and in the wider community. Mathematics still has an image problem, which teachers need to bear in mind when they are trying to come up with exciting, relevant, real-world problems. Topics like toxic waste, the destruction of the ozone layer, and the spread of an oil spill are all very tempting when a teacher is trying to add interest and relevance to the lesson, but mathematics already has such a negative image that it might be wise to be sparing with such material so that the subject does not become irretrievably associated with disaster (Clark 1990).

THE TEACHER

Most mathematics teachers have been trained on abstraction, and its pull remains strong. Recent high school statistics syllabi illustrate this well (New Zealand Ministry of Education 1991). Simulations abound, and usually these simulations are computer simulations.

Here is a subject area based in the real world, a subject to which female students are often drawn, and yet strong emphasis is placed on the abstract side of this very practical subject. Of course, in some destructive situations, simulations are important, but at the high school level it seems very difficult to justify extensive use of simulations if it is at the cost of losing the enthusiasm of those students who are attracted to statistics because of its relationship with data from the students' own environment.

Context counts certainly, but implementing change takes time. To include nappy (diaper) folding and infant-growth charts into the mathematics classroom seems straightforward enough, but teachers must tread carefully to avoid patronizing or mathematizing precious areas of people's lives that they do not wish to view in that way. As always in real ventures, a delicate balance has to be struck. The teacher's job is to get the balance right.

A single mathematical track is not suitable for all students, and many different paths to success are possible. In the 1989 New Zealand repeat of the SIMS study, a close look was taken at the classes that made high gains during the year (the students were tested in March and November). They were very diverse, and after all the possible sources for this growth in achievement were eliminated—school type, rural or urban location, percentage of Maori pupils, and socioeconomic status—what remained was the implication that for these students the year of teaching had really meant something. The teacher had made the difference.

When we ask the question that Abele (chapter 7 in this volume) puts to us, "What can we learn from these studies for classroom practice?" the answer seems clear. When mathematics problems are selected for assessment, the practice of skills, or the learning of concepts, the role of the teacher is vital because the choice of language and context for these problems is central to whether the problem is accessible to the learner.

REFERENCES

Clark, M. "Applications—Handle with Care." *Teaching Mathematics and its Application* 9 (1990): 33–34.

EQUALS. *Assessment Alternatives in Mathematics*. Berkeley: University of California at Berkeley, Lawrence Hall of Science, 1989.

Forbes, S., T. Blithe, M. Clark, and E. Robinson. *Mathematics for All*. Wellington: New Zealand Ministry of Education, 1989.

Goodman, K. *What's Whole in Whole Language?* Ontario: Scholastic-TAB Publication, 1986.

Howson, A. G., and S. Mellin-Olsen. "Social Norms and External Evaluation." In *Perspectives on Mathematics Education*, edited by B. Christiansen, A. G. Howson, and M. Otte. Dordrecht, Netherlands: Kluwer Academic Publishers, 1986.

New Zealand Ministry of Education. *Mathematics in the National Curriculum: Curriculum Statement for Mathematics Arising from the Achievement Initiative.* Wellington: Learning Media, 1991.

———. *Education Statistics of New Zealand 1991.* Wellington: Research and Statistics Division, Ministry of Education, 1992.

Reeves, N. "The Mathematics-Language Connection." In *Language in Mathematics*, edited by J. Bickmore-Brand. Carlton South, Victoria, Australia: Australian Reading Association, 1990.

Royal Society. *Girls and Mathematics: A Report by the Joint Mathematical Education Committee of the Royal Society and the Institute of Mathematics and Its Applications.* London: The Royal Society and The Institute of Mathematics and Its Applications, 1986.

Wily, H. "Women in Mathematics: Some Gender Differences from the IEA Survey in New Zealand." *New Zealand Mathematical Magazine* 23 (1986): 39–49.

19

Itineraries through Logic to Enhance Linguistic and Argumentative Skills

Giancarlo Navarra

University of Modena

O NE of the main problems that has to be faced daily by teachers is the difficulty the students have in oral and written expression. It is well known that linguistic competence is an essential condition for learning, quite apart from the disciplinary contexts, even the more technical ones that apparently require almost exclusively instrumental skills.

A complex aspect of linguistic competence lies in the ability to organize one's opinion in an exhaustive and communicable way; in other words, in the ability to *explain the stream of a reasoning* that starts from one or more premises and comes to a conclusion. To do that, it is necessary for the students to be *conscious* of their knowledge—apart from its scholastic or extrascholastic origin—and to be able to use it flexibly. It is important to ascertain how much the training and the organization of the student's knowledge can influence—positively or negatively—his or her skill in making internal connections. In other words, to what point is it possible for him or her to reconstruct his or her knowledge in front of new problematic situations if it has not been constructed from the beginning to favor flexibility?

To tackle these questions, I have conducted activities since 1986 in collaboration with a language arts teacher, Laura Buzzatti. We have constantly attended the lessons together. (The Italian education system allows this kind of cooperation within a particular arrangement of the school time called *tempo prolungato*.) These lessons were aimed at developing the potential of the linguistic skills in students aged 11 to 14. The teaching activity has focused on Logic, seen particularly as a *transversal area*—in comparison with linguistic and scientific-mathematical areas—and as a *reflection on language*.

After a short overview of the main objectives of this work, three examples are analyzed: To Argue and to Explain (11-year-olds), Nonbanal Reflection (12- to13-year-olds), and Deductions of a Shrewd Inspector (13- to 14-year-olds). The strategies followed by the students are illustrated. Finally, comments on some results of an analysis of the students' answers are presented.

AIMS OF THE ACTIVITY
AND METHOD OF WORK

The main aims of the work with students follow:

1. The improvement of linguistic skills through—
   - the development, by means of discussion, of critical thought, that is, the capacity of assessing the pertinence, the errors, and the potentialities of an utterance;
   - the order in organizing and communicating ideas respecting priorities, coherence among concepts, and properties;
   - the capacity to recognize communicative and metalinguistic levels.
2. The comprehension, on the one hand, of the richness connected with the ambiguity of the natural language and, on the other hand, of the power of a formalized language; in other words—
   - an identification of the differences between semantic and logical-syntactical aspects of speech;
   - the conquest of the concept of symbol;
   - the awareness of the relations between symbols and the rules that link them;
   - a competence in using instruments like truth tables.

Some of the more interesting aspects of this activity concern the active role of the students in directing discussions. In fact, original observations or points of view often lead discussions in directions not only unanticipated by the teachers but also highly sophisticated on the conceptual level for students of this age. At other times a debate is engaged between the two teachers, and students take part in a constructive way, as protagonists, while the continuous interplay between semantic and logical aspects creates premises to very interesting reflections.

Strategies are used that need different instruments: graphic instruments (Euler circles or Venn diagrams, Carrol's square, graphic trees), iconic instruments (transparencies, drawings, slides), manipulatives (Vygotsky-Dienes logical blocks, magnetic letters, playing cards, etc.). Linguistic instruments, however, are preferred: texts and stories, invented by the teachers, considered as logical problems (Malara 1989; Navarra 1992, 1993).

The large number of texts for every subject treated (connectives, quantifiers, "modus ponens," and other reasoning figures), proposed in sequence of growing linguistic complexity and logical difficulty; the flexibility in the running of sequences; the accurate diagnosis of arguments and linguistic difficulties; and the attention paid to the student-student interaction promote a progressive refinement of the critical skills of the students in assessing and comparing the adequacy of their own arguments and those of others.

The structure of this course allows the teachers to discover difficulties and abilities that, although well known to the teachers under some aspects and confirming what they already know about the students, may expand this knowledge in other directions. These other directions are usually less practiced in traditional teaching—for example, cross-comparisons, comments, and corrections of answers—with the aim of analyzing the validity of some of them and the linguistic clarity used to express them.

Such a standard of discussion can be associated with the *balance discussion*, carefully experimented with, and examined by other researchers (see, for example, Bartolini-Bussi [1991]). Of course, students in the class are not asked to examine all the answers, but just a selection of them—those that represent more clearly the different strategies. In this context, the discussion is planned and carried out with the essential support of the overhead projector, which is used to project the answers the students have transferred onto transparencies.

The lessons are usually verbalized to retain the richness of what is said during collective discussions, which otherwise would not be easily grasped. This enables a more careful reflection on the lessons already developed and the planning of the next lessons with more elements of judgment. These tasks are greatly simplified by the presence of both teachers in the class, which enables them to play, according to different situations, the roles of animator and of reporter.

ANALYSIS AND COMMENTARY
ON SOME MEANINGFUL STAGES

"To Argue and to Explain":
An Activity for the 11-Year-Old Students

Confronted with problematic situations with heavy logical content, students very often elaborate personal strategies to overcome cognitive obstacles. They change the given situations into situations that are not necessarily consistent with those given at the beginning but that are more able to be solved, in the students' view. It is of great interest to a teacher to examine these performances and to try to identify the table of logical inference that has been produced by the students.

Let us analyze, as an example, an initial experience conducted with 11-year-old students, before we begin the didactic-logical activity proper.

The text proposed is an adaptation of Smullyan's famous logical problem about knights and knaves (Smullyan 1978). The protagonists are a survivor from a wreck and the inhabitants of an island where the knights always tell the truth and the knaves always tell lies. The students, playing the part of the survivor, meet people. Because they do not know a priori who is a knight and who is not, students have to discover the identity of the various characters on the basis of the answers they give to the students' questions.

> Here comes the first inhabitant of the island. You look at him, but apparently he could be either a knight or a knave. So, without thinking, you ask, "Who are you, a knight or a knave?"

> What has the inhabitant of the island answered?

Let us examine the prototype answers:

Rosella: He answered, "I am a knight," but he knew he was not. He said so because he wanted the knaves to become more numerous than the knights; therefore, he knew or imagined that the knaves could become stronger than the knights.

Christian: I think he says he is a knight because if he is a knave he says he is a knight and if he is a knight he says he is a knight. My opinion is that the survivor should be able to understand it because if the inhabitant says he is a knave, he is a knight, and if he says he is a knight, he is a knight.

Elisa: He says, "I am a knight." I think he says so because if he is a knight, he tells the truth and he introduces himself as a knight, and if he is a knave he tells a lie and, instead of introducing himself as a knave, he presents himself as a knight.

It is clear that Rosella has given an imaginative answer, which can be assimilated into an *argument*, meaning that "every argument is personal, and addresses itself to people whose assent it tries to gain" (Perelman 1977, p. 791). Somehow, Rosella is trying to convince the audience to follow her logic, which is referring to situations and elaborations that are completely personal.

Elisa's answer, in contrast, looks somehow like a demonstration, although naive, as she deduces "with absolute certainty conclusions starting from premises, given according to rules previously formulated" (Perelman 1977, p. 791).

Christian gives an answer similar in structure to Elisa's, but he has not a complete logical control, and the second part (wrong) contradicts the first part (correct).

In this class, only three students give answers like the Elisa model; to various extents, the other sixteen follow the Rosella model. Almost everyone, however, while discussing the texts and confronting the two models, understands that only Elisa's answer is completely convincing, although the other ones may be funny, strange, or peculiar, they are less *sustainable*. Also, students notice that the explanations following the Elisa model are similar because they show evident analogies in the way the reasonings are structured; the others, on the contrary, are different from one another and can never be practically compared with one another. It is a difference similar to the one between *hypotheses* and *opinions*, elaborated by the students during the construction of simple mathematical models and also studied by the Research Group of Genova (Boero 1990).

Because such a situation emerged on other occasions also, it can be said that generally few students can recognize spontaneously the deductive connections among the elements of a problem and expound on them in a coherent form. As far as the other students are concerned, many of them, even when put in front of an individual activity, without the contribution of the collective discussion do not improve their performances in a significant way. It is important for them, however, to take part in the construction of the *test* of a reasoning and to grasp the processes through which the construction is realized. In our opinion their participation becomes much more significant if we consider it within the Vygotskian theory of the zone of proximal development (Vygotsky 1978).

Often, especially with linguistically weaker students (but not only with them), knowledge is confused because of the lack of ability to control the meaning of very common words that the teachers consider *known*.

A relevant example is that of 12-year-old students during a test stage on paraphrases of the connective "and": *also/too, on the contrary, however, but,* and so on. The aim was to lead the students to the identification of a logical characteristic common to words that have very different shades of meaning on the semantic level .

The test compared the possibility of a logical equivalence between two sentences:

- "I assure you," Aldo is saying, "the trout is a fish, while, *on the contrary,* the dolphin is a mammalian."
- "Absolutely not!" Cosimo answers back. "The trout is a fish; *however,* the dolphin is a mammalian!"

A below-average student openly declared his difficulties of interpretation: "Cosimo's answer could be right, but I cannot tell because I don't understand the difference between *however* and *on the contrary.*" It is not an easy task for teachers to evaluate answers of this sort once they are expressed and to choose the best strategies to address the difficulties indicated. A difficulty also arises because many times a student's confused explanation leads to the discovery of the existence of a problem but not to an appropriate therapy to solve it. This is especially true of students who have often already experienced six or seven years of predominantly frustrating schooling.

Certainly, however, logical linguistic activities, such as those proposed; the usual reflection about them; and particularly the activity of verbalization (oral and written) as constant didactic procedures can represent a very favorable pedagogical environment in this sense.

"Nonbanal Reflection": Activities for 12- to 13-Year-Old Students

Since the students are constantly obliged to reflect on their reasonings and on those of their schoolmates, the increase of the metacognitive level they are gradually using emerges as one of the most meaningful aspects of this study (see Schoenfeld [1985]). The desired purpose is to fulfill both the development of linguistic abilities and the refinement of students' capacity to find a self-direction of intellectual work (Boero 1990). This is considered necessary to improve their competence in the field of mathematics when, for example, they are consciously faced with problematic situations.

As an example of these last observations, a diary of the discussions treated in the classroom concerning this text is reported. The story follows:

Greedy Girls

Eloisa and Giovanna are admiring the window of the confectioner "Super Cake."

"Mmm, come on, let's go in!" says Eloisa. "First of all, I'm going to have a strawberry pastry and a chocolate shake."

"I can't see how you will do it with the little money we have. *Surely you can have the pastry _____ have/ing the shake,*" Giovanna observes while searching in her pockets.

The task: Fill the blank space with the connective that in your opinion Giovanna has used.

After a first stage of individual work the discussion produces the following suggestions:

(a) but not	8 students
(b) or	6 students
(c) without	1 student
(d) maybe	1 student
(e) and	1 student
(f) or else	1 student
(g) and then	1 student

The discussion brings the students to very interesting conclusions on the level of their comprehension of the meaning of the connectives.

Some of them observe that (e) and (g) are wrong because they contradict the context, in which Giovanna's perplexity concerning the scarcity of money is evident. It is so little that it is not enough for Eloisa's hypothetical bellyful, so it will not be possible "to have pastries *and then* the shake."

Choice (d), "maybe," suggests someone "with dots" ("*and ... maybe* to have a shake"), but it is still wrong for the same reason as (e) and (g): Giovanna is sure of what she is saying.

And now for the most meaningful contrast. Both (a), "you can have the pastry *but not* the shake!" and (c) "you can have the pastry *without* the shake!" are wrong because—according to the students—they create a hierarchy of preferences; (b) and (f), on the contrary, are right because they do not change the meaning of the sentence. Because it is impossible to know Eloisa's tastes, it is not possible to state whether she would prefer the pastry or the shake, as do (a) and (c) (there is the hierarchy!). In conclusion, (a) and (c) change the meaning of the text; (b) and (f) do not change it.

It is clear that during these reflections the contributions of the students are very differentiated. However, because of the strategy of constantly writing the students' observations and suggestions on the transparencies, the whole class is certainly busy in a performance that is very significant and highly refined at a conceptual level—comparisons, cross-corrections, changes of mind, self-criticism—in which the knowledge in question is developed both by the teachers and by the students.

"Deductions of a Shrewd Inspector": An Activity for 13- to 14-Year-Old Students

Between the ages of 11 and 14, students mature in terms of their ability to work competently in growing abstraction. The teachers are offered the possibility, by means of considering logical-linguistic situations, to start some activities that involve, even though at an early stage, higher symbolic and formal aspects. It is possible to replace reasonings, freely directed by the students, with others that imply a more structured knowledge of some subjects as, for example, *reasoning figures* and the research of their validity.

In the situation examined in the classroom, one of the aims is to compare the spontaneous methods with those of deductive logic to improve the organization and explanation of the students' thought and their capacity to argue and to favor the transition from a confused or undifferentiated analysis to a clear and coherent one.

The teachers decided to invent some thrillers, and a school for detective-students was begun to help Inspector Clouseau logically link clues and draw conclusions, in other words, solve the case.

This story was proposed in an examination at the end of this activity. The students were required to spot the guilty party's characteristics by using strategies and rules previously determined. For instance, Clouseau *is always honest.*

The Adventures of Inspector Clouseau on the Nile

Right at the end of the adventure described in "Death under the

Pyramids," the boat began the navigation again. A few days of calm have passed, when one night the inspector is suddenly awakened by the captain of the boat. "Inspector! Wake up! Someone has killed lawyer Fitzgerald! It's terrible!"

The poor inspector must interrupt his sleep and set to work again. As usual the inquiries are complicated, and at the end of the day Clouseau is reflecting on the depositions he has heard.

"Well, the guilty person could be either one of Fitzgerald's colleagues or one of the German tourists; I'm not certain yet … but of one thing I'm sure: the witness who told me that the killer is a lawyer and a short person is a liar! And I've understood that even Mr. Ross, who assured me that the killer is not American, lies … then, let's see: considering that.…"

At this point in the inquiries, the suspicions of Inspector Clouseau are still on all the tourists, nobody excluded.

Before examining the results of the test, let us analyze briefly the difficulties the students must overcome:

1. To identify the three pieces of circumstantial evidence:

   (a) "The guilty person could be either one of Fitzgerald's colleagues or one of the German tourists."

   (b) "The witness who told me that the killer is a lawyer and a short person is a liar!"

   (c) "Mr. Ross, who assured me that the killer is not American, lies."

2. To explain, by writing in a clear way, the connections among the pieces of evidence so as to identify the killer among the guests on the boat. To do that, it is necessary to identify the logical aspect of the three statements on the basis of the presence of connectives and truth values:

   (a) The first piece of evidence contains an inclusive disjunction (either … or); its general truth value is *true*—the inspector is sure of what he is saying, and we can trust him. (As we have said before, Clouseau is always honest.)

   (b) The second piece of evidence contains a conjunction (and), and its truth value is *false* (the witness is a liar!).

   (c) The third piece of evidence contains just one statement, and its truth value is *false* (Mr. Ross lies).

Referring to truth tables and selecting the possibilities compatible with the relative truth values of the statements, the students can recreate Clouseau's reasoning.

According to the quality of the explanations (relative to this text and to other logical problems), the students can be divided into three groups (relative to a pentenary distribution A, B, C, D, E). It seems possible to draw the partial conclusion that clearness in the control of the logical contents and clearness in the linguistic expression are strictly linked to each other.

The A/B group

The students show the highest linguistic competence, carrying out the analysis of the stories correctly and understanding clearly their logical contents. Also, the synthesis is generally arranged coherently (although sometimes it may be too hurried) through an organized selection of the sentences, which allows the conclusion to be reached.

The logical mistakes are of little importance and are always easy to find, and the students are able to correct them consciously.

Most of the students organize a discursive answer—sometimes too complex, even prolix, to the detriment of clearness and in other instances, more schematic. Some even make an improper stenographic use of logical symbols , which teachers of mathematics know very well is a stereotype very often used by students while working with algebraic symbols.

The C group

The students commit inaccuracies on various levels of the analysis as they interpret, in a partially incorrect way, the connectives or the truth indicators. Others hastily and superficially reach a conclusion that even when correct in its substance is without a clear articulation of the synthesis.

Mistakes cannot always be easily found. Sometimes they are concentrated, sometimes they are sprinkled throughout the course of the reasoning. The teachers, therefore, have to employ different strategies and need a long time to discover the causes that have induced these mistakes.

The text is often inaccurate, and it is frequently composed of schematic parts, linked by short discursive parts; the whole text reveals a general linguistic poverty.

The logical and linguistic levels are in part confused. The students cannot overcome a kind of detachment from the logical sense of what they are doing. In a certain sense they remain outsiders to it.

To take advantage of the connections among the situations they are examining, they need the others to listen to their arguments and to discover in them a posteriori the clearness they have only incompletely. Often they acquire a reasonable control of the instruments, but without the sense of the context in which to use them.

The D/E group

These students show misconceptions related both to the identification and to the interpretation of the logical contents of the text. The reasonings are linked again to a personal logic (the Rosella model)—in spite of the two years that have gone by since the activity about knights and knaves—and they are built tentatively. The students concentrate their efforts only on the analysis and interrupt it before having drawn some conclusion, or they get to a wrong conclusion in an incoherent way.

The mistakes are many and diffused, and the strategies to correct them, at least in part, are more complex to define and require individualized interventions.

There are no series of problems. Every problem appears as a special case, so it must be faced with improvised instruments and strategies. The students are hardly aware either of their knowledge or of the linguistic means to express it properly.

The differences among many D/E students' texts and those of others are salient. They seem worried, above all, that they will not be able to convince the teacher. For this reason, they use—in a confusing, redundant, and stereotyped way—every kind of language they have learned during the activity, making a lot of mistakes. Evidently the didactic contract is not clear to them (Brousseau 1986).

A/B students, on the contrary, are able to be more selective, concentrating only on the necessary information from which they create synthetic texts.

The divergent analysis of texts, followed by the explicit research of the reasons that one text is objectively clearer than another, and the need to take a stand in the critical assessment of the texts are situations that have a high didactic value and allow the students to pinpoint some of the main reasons for clear reasoning: the information is correctly used—neither superabundant nor insufficient; the parts of the reasoning are positioned in a very comprehensible order (from the premises to the conclusion); more explanatory clearness is apparent; no confusion exists among different languages; the correct syntax, spelling, and grammar are used; on the whole, a higher formal order is present; and so on.

Moreover, two styles of work emerge: to anticipate immediately the conclusion (chosen by the majority) or to put it at the end of the reasoning process. One method is not necessarily better than the other. During an introductory period of study in this topic it may be more advisable to follow the second way, which requires more methodical work (first I analyze the data, *then* I can draw the conclusions) and also to avoid the mania of the result of many students, which, at least at the beginning, could be mitigated by the duty to follow, more strictly, a plan of work.

As we have previously stated, it is not easy for weaker students to discuss their solutions, particularly when they are not used to explaining and arguing in class. At this point, it must be admitted that often no homogeneity can be found among the teachers—even of mathematics (see Pimm [1987])—when they have to organize educational strategies in the linguistic field; consequently, not all of them dedicate time to such a significant activity. They prefer to delegate this work to their colleagues in the language arts department. Experience (not just that gained from this study) confirms, however, the efficacy of a working method that favors metacognition and linguistic reflection and its high potentiality—even psychological—in relation to the acquisition by the students of a greater confidence in themselves, in their schoolmates, and in their teachers.

FINAL REMARKS

In conclusion, some open aspects of this work should be stressed.

A first remark concerns the analysis of different and articulate reasons that could be at the basis of the attitudes of students following the Rosella model (more generally, the D/E students), namely, at the basis of their generalized difficulties in front of logical problematic situations, for example, the social-cultural conditioning or a doubtful or unsuccessful didactic contract.

Another remark concerns the difficulty of evaluating the relapses that activities like this can induce in the general linguistic competence of the students outside such an unusual work. These activities tend so precisely to the understanding, according to the situations, of differences or logical-linguistic analogies even when not very evident.

This is difficult too because in the normal didactic practice, the opportunity to reproduce situations so favorable to linguistic reflection seldom occurs; moreover, not every teacher pays so much attention to written and oral verbalization.

It is equally complex to discover clearly the connections between the development of linguistic competence and the improvement of mathematical knowledge, but many researchers (e.g., Laborde [1990]) acknowledge the efficacy of verbalization and of discussions among students, even while they underline the need for becoming better acquainted with the dynamics of the linguistic interaction in the class and for knowledge of more propitious didactical contexts.

This work reveals three opportunities for its accomplishment, even within the syllabi established by the Italian government: a significant flexibility in the achievement of the aims of the program; the two-hour lessons every week undertaken by the two teachers working together constantly; and a willingness to do a regular comparison, which could be defined as epistemological, between the two disciplines of mathematics and language arts.

This enables us to point particularly to the quality more than to the quantity of the knowledge—without excessive problems about the time necessary for finding it. These are probably infrequent aspects in the ordinary school routine.

Indeed, it is necessary to have long periods of time at one's disposal because it is the class as a whole that conditions the development of the activity and guides the teachers in one or another possible direction. This compels them to direct attention toward the large variety of answers offered by the students, to work out (often at a very short notice) an operating schedule of future activities. Consequently, they may propose situations that are more suitable for clarifying or adjusting concepts that caused problems for the students and also because of the frequent ambiguities implicit in the natural language.

In other words, it is a matter of a kind of work in which a circular relation develops between an accurate analysis of the situation to be proposed and of its possible expansions and the critical interpretation of the student's attitudes and of the protocols created by them *in itinere*.

REFERENCES

Bartolini Bussi, M. "Social Interaction and Mathematical Knowledge." In *Proceedings of the Fifteenth Conference of the International Group for the Psychology of Mathematics Education*, Vol. 1, pp. 1–17. Assisi, Italy, 1991.

Boero, P. "Allievi con difficoltà di apprendimento: Che fare?" *L'insegnamento della matematica e delle scienze integrate* 13 (1990): 1171–90.

Brousseau, G. "Fondements et méthodes de la didactique des mathématiques." *Recherches en didactique des mathématiques* 7 (1986): 33–115.

Laborde, C. "Language and Mathematics." In *Mathematics and Cognition: A Research Synthesis by the International Group for the Psychology of Mathematics Education, an ICMI Study*, edited by P. Nesher and J. Kilpatrick, pp. 53–67. Cambridge: Cambridge University Press, 1990.

Malara, N. "Implicazione e modus ponens: Sintesi di una esperienza didattica realizzata in una seconda media." In *Atti degli incontri di logica matematica, XII, Roma, 6–9 aprile 1988*, edited by M. Barra and A. Zanardo, pp. 245–53. Padova: Grafica GSE, 1989.

Navarra, G. "Il signor Kappa allo zoo, riflessioni su un'attività didattica sulla comprensione dei valori di verità dei connettivi 'E', 'E/O', 'O'." *Scuola e didattica* 13 (1992): 83–88.

―――. "Un mazzo di fiori alquanto pericoloso, itinerari didattici attorno alle Leggi di de Morgan." *Scuola e didattica* 9 (1993): 36–40.

Perelman, C. "On 'Argomentazione.'" In *Enciclopedia* Vol. 21, pp. 791–823. Turin: Einaudi, 1977.

Pimm, D. *Speaking Mathematically: Communication in the Mathematics Classroom*. London: Routledge & Kegan Paul, 1987.

Schoenfeld, A. "Metacognitive and Epistemological Issues in Mathematical Understanding." In *Teaching and Learning Mathematical Problem Solving: Multiple Research Perspectives*, edited by E. A. Silver, pp. 361–79. Hillsdale, N.J.: Lawrence Erlbaum Associates, 1985.

Smullyan, R. *What Is the Name of This Book? The Riddle of Dracula and Other Logical Puzzles*. Englewood Cliffs, N.J.: Prentice-Hall, 1978.

Vygotsky L. S. *Mind in Society: The Development of Higher Psychological Processes*. Cambridge: Harvard University Press, 1978.

20

Communication in a Secondary Mathematics Classroom: Some Images

Judith Fonzi

University of Rochester

Constance Smith

State University of New York at Geneseo

THIS chapter gives an account of several days in a rather unusual high school mathematics class. The goal in doing so is twofold. First of all, it should provide a rich image of the variety of ways in which communication can occur in mathematics instruction. Second, in light of this shared example, it should stimulate questions and discussion about the roles that communication can play in learning and teaching mathematics.

Such seemingly simple questions as, In what ways can communication take place in a mathematics classroom? Who can be involved in the communication? or What could be the content of such communications? are not so simple when viewed in the context of an actual class. This is especially so if we look at how communication develops over a considerable period of time and in a mathematics class based on an inquiry approach (Borasi 1992; Siegel and Borasi 1994) that emphasizes the construction of meaning through active, social, and purposeful experiences.

The "Reading to Learn Mathematics for Critical Thinking" project was supported, in part, by a grant from the National Science Foundation (Award #MDR—8850548). The opinions and conclusions reported in this chapter, however, are solely the authors'.

More specifically, it is hoped that the instructional experience reported in this chapter will provide a concrete context to address some of the following questions: If communication is considered as more that just verbal conversation, what other modes of communication are available in a classroom environment? How can these modes of communication be used in mathematics instruction? What can students talk about in a mathematics class, besides specific mathematical concepts or techniques? What would communication about the nature of mathematics, the process of doing mathematics, or feelings about what is being learned look like? What happens when students take a more central role in the communication taking place in the mathematics classroom?

Given the goals as stated previously, the centerpiece of the chapter is a narrative report of an eight-day unit that developed around the distinction between analog and analytical thinking in mathematics. This experience occurred within a course on Math Connections offered to high school students in the United States. Given the space constraints, the narrative mostly summarizes each key event constituting this experience, with a few excerpts of the actual dialogue interspersed to give a better sense of the communication taking place, and some commentary to highlight some interesting points about such communication. This narrative also enables the introduction in context of the various categories that could be identified in the systematic analysis of this experience with respect to who was involved in the communication, what the communication was about, and the mode of communication used. For the sake of brevity and clarity, not all instances of communication are codified in this narrative; however, a summary of the configurations, contents, and modes of communication that were identified in each key event can be found in table 20.1. A brief discussion of the main questions posed in this introduction concludes the chapter.

LOOKING AT COMMUNICATION IN ACTION IN AN UNUSUAL MATHEMATICS CLASS

Instructional Context

The experience presented here was part of a one-semester course, Math Connections, designed and taught by Judith Fonzi. This course took place at the School Without Walls, a public, urban, alternative

Table 20.1

Summary of the Modes, Configurations, and Contents of the Communications for Each Key Event

		1	2	3	4	5	6	7	8	9
Modes	Reading	•	•		•			•		•
	Writing	•	•		•	•	•	•		•
	Telling/showing	•	•				•	•	•	
	Modeling/demonstrating	•	•		•			•		
	Discussing	•	•	•	•	•		•		•
	Drawing	•	•					•		
	Viewing/observing						•			
Configurations	Student-student	•	•	•	•	•	•	•	•	•
	Student self	•	•					•		
	Student-teacher				•	•				
	Teacher-student	•	•		•		•			•
Contents	Technical mathematics	•	•				•	•	•	
	Nature of mathematics	•	•	•	•	•	•	•		•
	Process of doing mathematics	•		•				•	•	•
	How one is learning	•	•		•					
	Feelings about learning	•	•		•	•				•

Event 1: Two students read, and share their understanding of, the essay "Analog and Analytic Mathematics."

Event 2: Other members of the class become actively involved in analyzing the given examples of analog and analytic thinking to better understand these two approaches.

Event 3: The class generates and examines new examples in their continued efforts to understand the nature of analog and analytic thinking.

Event 4: Pulling all the information together to articulate clearly the distinctions between an analog and an analytic approach

Event 5: Reflecting on this part of the course and looking ahead

Event 6: Setting the stage for an inquiry in geometry to further enhance students' understanding of the nature of mathematics

Event 7: Small groups create their own theorem about two different figures that have the same area and prove it, using both an analog and an analytic approach in the process (Group 2 only).

Event 8: The groups synthesize their work and communicate the results of their inquiry to the whole class.

Event 9: Completing the inquiry by creating a context for their theorem secondary school characterized by the goal of developing independent learners through a focus on learning how to learn. The main goals of this course were *(a)* to broaden students' conceptions of mathematics, so as to challenge the commonly held notion of mathematics as a set of rules and procedures and to enable them to see mathematics as a way of thinking (Bishop 1988) and a human construction and *(b)* to support the school's overarching goal to help students learn how to learn. The theme of looking for connections between mathematics and other fields was chosen as a context within which to work toward these goals. The design of the course was also influenced by the teacher's participation as a teacher-researcher in the "Reading to Learn Mathematics for Critical Thinking" research project, an interdisciplinary project exploring the role that reading could play in mathematics instruction (Siegel, Borasi, and Smith 1989; Borasi and Siegel 1994; Siegel and Fonzi 1995; Siegel et al. 1996).

The experience reported here occurred in the second half of the course, when the students independently read and then "taught" to the rest of the class several essays from Davis and Hersh's *The Mathematical Experience* (1981). These essays had been chosen because they focused on the social and historical development of mathematics or on various approaches to doing mathematics and thus could validate and push even further students' broadening views of mathematics. Each student was expected to choose and read one essay. They were to select and use a reading strategy from among several transactional reading strategies (Siegel 1984; Harste and Short, with Burke 1988) that they had learned in previous parts of this course and then share the sense they made of their essay with the class. The example begins with the sharing of the final essay, "Analog and Analytic Mathematics," and continues through an inquiry that was developed to further support students' efforts to make sense of these two approaches to doing mathematics. We provide the following excerpt from the essay so that the reader can better appreciate the style of the texts being read as well as the content of what the students in our example had read.

> Analog mathematizing is sometimes easy, can be accomplished rapidly, and may make use of none, or very few, of the abstract symbolic structures of "school" mathematics. It can be done to some extent by almost everyone who operates in a world of spatial relationships and everyday technology.... In analytic mathematics, the symbolic material predominates. It is almost always hard to do. It is time consuming. It is fatiguing. It requires special training.... Analytic mathematics is perfomed only by

very few people.... [*The authors further illustrate the nature of these two approaches through a series of examples contrasting an analog and an analytic approach to specific problems.*] **Problem:** How much liquid is in this beaker? *Analog solution:* Pour the liquid into a graduated measure and read off the volume directly. *Analytic solution:* Apply the formula for the volume of a conical frustrum. Measure the relevant linear dimensions and then compute. (Davis and Hersh 1981, pp. 302–4)

THE INSTRUCTIONAL EXPERIENCE

Event 1: Two students read and share their understanding of the essay "Analog and Analytic Mathematics"

Reading the essay

Tim and Char each read the essay for homework, to make sense of its content and to prepare to share this sense with the class. (We consider this an example of communication that involved the teacher communicating to students (configuration), was about the nature of mathematics (content), and occurred through the use of reading (mode)). Both Tim and Char wrote notes to themselves about the content of the essay, their responses to the reading, or both (*students communicating to self—about the nature of mathematics—through writing*). Before completing the reading, Char also stopped and drew a picture, which she felt represented an analog and an analytic solution to the problem of "how to go to Tanya's (her friend) house" (*student communicating to self—about technical mathematics—through drawing*). She tested the accuracy of the example by reading the rest of the essay to see if it continued to fit her interpretation of the two approaches.

Notice that the teacher has orchestrated an experience that will require the students to communicate with each other about very specific issues. She has chosen to convey to students certain information about these issues by asking them to read a text written by someone else. Since the students then had to communicate this information to others, they, too, had to think explicitly about what information to share and what mode of communication to use to convey this information.

Sharing what they learned

Since Tim said that he did not really understand the essay, the class decided that Char should explain what she understood from it and then Tim should give his perspective. Char thus began:

Char: Umm ... I think I [understand it]. I'm kind of having second doubts, but maybe.... I did "Analog and Analytic Mathematics." And I read half of it, and I did the cards, and I outlined my paragraphs and what I was going to put on what cards and ... but I read half of it and all of a sudden a light-bulb went off. And so I drew this neat little picture (*student communicating to students—about how she is learning—through telling*).

As Char is speaking, the teacher is in the background writing on the chalkboard. She is setting up a chart as a way to record the information shared by the students about the distinctions between analog and analytic approaches to doing mathematics (*teacher communicating to students—about how she is learning—through modeling*).

Char: (*Showing her picture representing two routes from home to her friend's house, a straight, direct path marked "analog" and a winding path marked "analytic"*) What I did.... It's a really weird drawing now, but it was really cool then. Analog its ... for both of these—it doesn't matter what you are doing—you get to the same point. Okay. So you are starting at the same point and you are getting to the same point but it is different how you get there. So I draw this ... I thought of ... like when I walk over to Tanya's house. Do I walk the fastest way? Or ... but I still get there. Or do I kind of like ... go around the block just because I like to walk down that street? But I still get there. And that's what.... I finished reading the paper, kind of, and it fit. And that's all I did and then I put it away and said I was done with my math homework (*student communicating to students—about the nature of mathematics—through showing and telling*).

Skeptical of what they heard, Shellie and Krista then questioned Char's approach and her explanation of the content of the essay. Char responded that this was her understanding of the material and then offered another example: "Analog ... it's kind of like two plus one equals three ... and analytic is one plus one plus one." After hearing this additional example, Shellie and Krista announced that they could now accept Char's approach to the reading task and, furthermore, that they now understood the concept she was teaching (*students communicating to students—about their feelings about what they are learning—through the initiation of an ongoing discussion*).

At this point, Tim joined the discussion with his own interpretation of the essay, which seems to contradict Char's:

> Tim: (*after referring to his notes*) I saw analytic as the simple way of having things done. Like her simple way of getting to Tanya's house. And I put analog as in the part ... computing ... maybe, analytic is like ... us ... what we are doing instead of having books and stuff. It's thought that's doing it. (*He is referring to the process of teaching and learning in the Math Connections course*) And with analog it is writing down on paper and computing and stuff like that (*student communicating to students—about the nature of mathematics and about our process of doing mathematics through telling*).

What is interesting and unusual in the communication taking place here is, first, that a student is sharing how she made sense of the essay and, second, that other students were the ones questioning the validity of her interpretation and her process. It is also important to notice that by allowing Tim the opportunity to hear someone else's interpretation, he is able to use it as a springboard for articulating his own interpretation. Up to this point, the teacher's role has been to facilitate the sharing by providing a structure to help the students make sense of the information; she has not contributed to the content of the conversation or questioned its accuracy.

Analyzing and synthesizing the information shared

In an effort to clarify Tim's and Char's apparently different interpretations and to enrich the whole class's understanding of the analog and analytic approaches to doing mathematics, the teacher then decided to involve all the students in sorting out the information that had been presented, determining what remained unclear, and stating what still needed to be done. She did so by returning to the chart she had begun to create when Char explained her interpretation of analog and analytic thinking and by insisting, from her own perspective as a learner, that she needed to fill in the chart with the examples and characteristics that had been shared so far. To fill it in, she urged Tim to repeat his distinctions for each approach. But as he did so, he began to change his mind and to identify analytic as the hardest approach and as computing. Neither Tim nor the other students who tried to help him could make sense of his notes so as to confirm his most recent interpretations. Thus the teacher, acting once again as a member of this community of learners, publicly enacted the next step in her learning process. She said aloud that she did not feel that she had been given enough information or the right kind of information to make sense of the two approaches to doing mathematics. She further elaborated her position by outlining the information and the sense she had made thus far and explaining why she did not feel that she understood what the two methods were.

In response to these concerns, Tim began to talk about a specific example referred to in the text: How much water is in a glass? He said that the analytic solution would be to measure the circumference, base, and height with a ruler and do on paper the computations for the volume of the glass. In contrast, the analog solution would be to pour the water into a premarked measuring cup. Tim characterized this second method as "Fast!"

Since class time was almost over, the teacher recorded Tim's new contributions on her chart without further discussion. At this point, since the discussion had focused primarily on the notion that an analog approach is "easier" but that both approaches would end in the same place, the teacher was not satisfied that the class (or even Tim and Char) had been able to construct a clear and accurate understanding of the distinctions between the two approaches to doing mathematics or of when, why, or how one would select one method over the other. However, not wanting to usurp the students' control of the experience, she suggested that they tell Tim and Char what they thought they still needed to know from them to make sense of the different approaches to doing mathematics. The students requested a definition of each method. (This entire activity is a good example of the *teacher communicating to the students—about her learning process, about the nature of mathematics, and about her feelings about what she is learning—through discussions and modeling.*)

It is important to note here that even when the teacher did become actively involved in the discussion, she did not take the usual role of fixing the errors and clarifying the information. Instead, because of her genuine position as a learner, she was able to express her confusion about what she had heard. This honest response acted as an invitation to other members of the community to become involved in the discussion by communicating their own understandings. The teacher was then able to communicate through her actions (modeling) a strategy for what to do when confused.

Event 2: Other members of the class become actively involved in analyzing the given examples of analog and analytic thinking so as to better understand these two approaches

Revisiting the previous examples

At the beginning of the next class, a student focused the discussion by restating the questions that Tim and Char had been asked to address: How do the analog and analytic approaches work? What is the difference between the two? Since Char was absent, Tim tried to address these questions by revisiting the example of measuring the water in a glass, drawing a picture as he spoke. Once again he contrasted an analogic solution, pouring the water into a measuring cup, with "the hard way, by measuring the base and the height and the weight and, you know, the circumference and stuff ... that way is analytical." At that point other students began to publicly take up the task of analyzing the information presented. Jo and Eric responded to Tim's analytic approach with the questions, "Could you explain that?" and "Why is it analytical ?" Tim answered by saying, "It means doing everything by the book. You know, breaking out the book." Hoping for some help clarifying this issue, Tim then immediately responded to Shellie's raised hand.

Shellie, however, did not want to respond to what Tim had just said but rather to what the teacher was writing on the chart, begun the previous day, about analog and analytic approaches as a result of Tim's example. She said that she was thinking about what the teacher was writing on the chart and wanted to share her opinion. Shellie said she liked Char's example (from the previous day) of analog as two plus one equals three and of analytic as one plus one plus one, adding, "That helped me understand it." Prompted by Shellie's comment, Margie (a university collaborator), Shellie, Tim, and Krista became involved in a discussion in which Margie spontaneously facilitated a public analysis of the example from her genuine position as a newcomer to the discussion. The discussion allowed the class to revisit old examples and identify some additional characteristics of the two approaches under discussion.

> *Margie:* Shellie, What's the difference there? Can I ask? I didn't understand that. I was gone last week so....
>
> *Shellie:* Char did the same reading as Tim. She brought up that umm ... the way they are different is analog which is easier is like two plus one equals three and analytic is harder which you've got to go to a long route to get to it which is one plus one plus one equals three. And that really helped me understand it.

Margie: Oh. Now was she using that ... those problems.... What's important there? The two plus one or the idea that it's an easier approach?

Shellie: Yeah. It's an easier approach.

Margie: So, the general sense is that the analog ... what that example shows is that one is a shorter route and easier. But you don't mean ... they're not really talking about addition problems, or are you?

Shellie: Well, it could be something else.

Tim: It may not even be easier, but just less time. Maybe less thinking, too.

Shellie: You are going to come to the same outcome.

...

Tim: Maybe even ... even less ... less of a thought, too. Because I mean ... there's more thought in this than there is in this. (*He refers first to the formula and calculations for the volume problem and then to the measuring cup.*)

Krista: She (Char) also made the statement of going to Tanya's house. You could go the fast route or shortcuts or something or you could go all around your favorite streets but you get there the same way only one is longer than the other.

This discussion is a good example of the *teacher* (in this case, the university collaborator) *communicating to students—about the nature of mathematics—through modeling and discussion.*

At this point, believing that the volume example was a good choice of problem to show the different approaches, and a bit concerned that no one objected to Tim's inaccurate explanation of the computation, the classroom teacher entered this discussion for the first time by explicitly sharing her own needs as a learner in this community. She told the class that she was trying to compare Tim's analog and analytic approaches to finding the amount of water in the glass but was having difficulty because she did not quite understand the computations in his analytic approach.

Analyzing the volume example by carrying out the analytic approach

The whole community then became engaged in constructing a process for computing the volume of a glass of water. Through a lengthy conversation that exposed students' past experiences, and through spontaneous research in some of the classroom texts, the class discussed what volume was, decided what figure would be suitable for the glass (a cylinder), and generated possible formulas for finding the volume.

While the class was deeply involved in identifying a formula for finding the area of the circle, some students indicated that they had lost sight of the purpose for this discussion through such comments as, "I thought this wasn't supposed to be a traditional math class," and "I just wanted to say what does this all have to do with the two approaches?" It was another student, Krista, who took the initiative to address these comments. She carefully reconstructed the events that led to finding the volume of the cylinder and then reviewed the procedures that brought them to their current point in the process of finding the volume. The teacher added that she felt that it was important to understand Tim's example because it was being used as the primary example to distinguish the two approaches to doing mathematics (*students communicating to students—about their feelings about what they are learning and how they are learning—through discussion*).

Ultimately the class was able to establish a formula for finding the volume of a cylinder, reconcile it with the texts' formulas, and then analyze the example for what it contributed to their understanding of the analog and analytic approaches to doing mathematics (*students communicating to students—about technical mathematics—through discussion*).

It is interesting to notice how the nature of the communication changed from Event 1 to Event 2. This is no longer a presentation by specific students on the essay they read. It has become a community effort to understand what the two ways of thinking discussed in the essay really are. The whole class has become involved in analyzing examples to identify characteristics of the two approaches. In addition to the change in the nature of the communication, the students are also taking up some new roles, which were modeled by the teacher in Event 1. Students are asking one another to clarify their examples, they are seizing opportunities to share what helped them understand, and they are intervening when they become confused about what or why they are doing something. It is this type of genuine communication among the learners about their sense making that sustains the students' interest in the topic and deepens the level of the discussions.

Event 3: The class generates and examines new examples in their continuing effort to understand the nature of analog and analytic thinking

A student tests her understanding of the two approaches by offering a new example

Shellie tested her own understanding by asking the community to verify an example she offered. The students, using the process established earlier in the discussion, analyzed her example and confirmed it on several previously established grounds.

Identifying more characteristics of analog and analytic thinking through new examples

Following Shellie's lead, the teacher offered another problem for the community to consider. The problem of deciding when to cross a street had been discussed during the sharing of a previous essay about conscious and unconscious mathematics (Davis and Hersh 1981), and the teacher now asked what would be an analogical and an analytical solution in this case. This example was intended to broaden students' appreciation of each of the approaches, by moving beyond easy versus hard, to consider issues of context, such as which approach is more appropriate—or even feasible!—given a particular problem. Students identified various solutions to the problem of deciding when to cross a street and determined that an analog approach made more sense in this situation. They went on to describe several new problem situations in which an analytic approach would make more sense. Thus the class concluded that one cannot decide a priori which of the two approaches is better but rather that it depends on the specific problem considered.

At this point it is interesting to look at the stages of this experience to see the role being played by the different modes, configurations, and contents of the communications. The experience began as a presentation—two students telling what they learned about analog and analytic approaches to doing mathematics from reading the essay. However, as a result of their sharing the tentativeness of their understanding and their willingness to accommodate each other's interpretation within their own understanding, the class saw a need to examine further the characteristics of the presenters' examples. Although this was a dynamic and generative investigation that produced some additional qualities and characteristics of analog and analytic approaches to doing mathematics, the tone of the conversation continued to be one of trying to understand the ready-made facts of the essay. As a result of examining the original examples, the class developed a shared sense of the nature of analog and analytic thinking and generated an initial list of the characteristics of each approach. Finally, students were then able to use this newly developed community understanding to begin to generate their own unique examples of analog and analytic solutions to problems. At this point the teacher was able to push the discussion to the next stage, at which the community was no longer trying to understand the facts from the original examples but rather could extend beyond these to construct its own ideas about analog and analytic approaches to doing mathematics.

Event 4: Pulling all the information together to articulate clearly the distinctions between an analog and an analytic approach

Taking stock of their understanding of the two approaches

The teacher now facilitated a more narrowly focused discussion to ensure that the students had been able to untangle the information and could clearly identify the characteristics of, and main differences between, analog and analytic approaches to doing mathematics. To support and affirm what the students had found, she initiated the discussion by sharing the following notes, which she had taken when she read the essay:

Analog	Analytic
• Sometimes easy	• Symbolic material dominates
• Accomplished rapidly	• Almost always hard
• To some extent everyone can do it	• Time-consuming
• Uses few or no abstract symbolic structures of "school math"	• Requires special training

As she read her list aloud, all the members of the class (even those who had not previously been very verbal) elaborated the meaning of each item. The students finished this discussion by reflecting on, and making connections to, the kinds of work they had done in previous parts of this course. They generated the following list of items, which included examples from their previous work:

Analog	*Analytic*
• Much of unconscious math	• Conscious math
• Measure	• $V = b \times h$ (their formula for volume of a cylinder where b represents the area of the base)
• (Using one's) experience	
• Understanding how a car engine works by developing a general picture of how the parts of a car engine go together	• Understanding how a car engine works by developing a formula for gas mixtures, a formula for calculating timing, actual tolerances of every engine part

Notice that even though the class has recapped and regrouped throughout this experience, the free-flowing nature of the communications in an inquiry-oriented classroom made this formal cleaning-up, organizing, and reflecting on the information a necessary step. It afforded the students an opportunity to slow down and integrate what they had learned.

Event 5: Reflecting on this part of the course and looking ahead

At this point the class moved on to a reflective discussion of the entire set of readings from *The Mathematical Experience*. After first generating their ideas in writing, the students shared their thoughts and feelings about the effectiveness of the reading strategies they had used and their potential for future use and about the connections among the essays and to the theme of the course. During this discussion the students noted that one could approach the study of mathematics from a historical perspective, a social perspective, or a "how to do mathematics" perspective. As the class period was ending, the teacher told the students that they would be moving on to study a specific mathematical concept by using all these approaches.

Notice how the conversation has shifted to a different level, from understanding the details of a specific issue about the nature of mathematics discussed in a single essay to making the connections among issues that have emerged throughout the course to develop a global sense of the nature of mathematics. Notice also that this conversation acts as a bridge between the past experiences and the next one.

Event 6: Setting the stage for an inquiry in geometry to further enhance students' understanding of the nature of mathematics

In an effort to further the students' understanding of the multiple approaches to studying mathematics, and to apply them specifically to what the students called *school math* or *traditional math*, the teacher then orchestrated an inquiry about the concept of area by using the Pythagorean theorem as a model. The students watched a video, *The Theorem of Pythagoras* (Project Mathematics! 1988), which discusses this fundamental theorem by giving a historical account of the many different times, places, and ways the theorem was discovered or proved. Through the use of computer graphics, this video shows, without labeling them as such, both analog and analytic approaches to proving the theorem. The students were told that they were not expected to fully make sense of the technical aspects of the mathematics from this video; rather, the purpose for viewing the video was to get a sense of how a specific theorem was studied from multiple perspectives. They were asked to identify, and jot down, examples of analog and analytic approaches to the theorem and to write down any general comments about the approaches and what they looked like (*teacher communicating to students—about technical mathematics and the nature of mathematics—through viewing and observing*).

In the next class, as the students shared the ideas about the nature of analog and analytic approaches that they had recorded while watching the video, the teacher created a list that integrated all these ideas. On the basis of this list and their previous discussions, students were then asked to identify themselves as someone who usually takes an analog approach or someone who usually takes an analytic approach and then to form heterogeneous small groups with respect to their predominant approach. The groups were given the task "to create a theorem about two different figures that have the same area and prove it, using both an analog and an analytic approach in the process."

It is interesting to note that even though the teacher has orchestrated a small-group inquiry experience that, by its very structure, requires students to communicate with one another to some extent, she continues to orchestrate additional uses of communication in a variety of ways for a variety of purposes. In this situation, she has again used an outside source—a video—to convey certain messages as a catalyst for the experience. However, this time it was intended to act as a model for the inquiry in which the students would engage rather than as the source of specific information to be learned. In addition, the public sharing of students' observations about the use of analog and analytic approaches in the video generated a list of strategies that not only allowed them to determine appropriate group members but could also be useful later as the students carry out their inquiry.

Event 7: Small groups create their own theorem about two different figures that have the same area and prove it, using both an analog and an analytic approach in the process

Each group was given plain paper, lined paper, scissors, and a sheet with some formulas for the area of some simple figures. In addition, the teacher reminded the students that the video and all the classroom resources (e.g., texts, materials) were available. The inquiry took most of two class periods, during which each group had complete autonomy over how it completed the task. What follows is a report on what happened within one of the groups, to illustrate the kind of communication that took place during this type of activity.

Group 2—making sense of the task

Several members of this group immediately offered their individual ideas in response to the given problem. One girl drew two semicircles and a circle and said, "Take a circle and cut it in half and then one half plus the other is the circle." Another student said that they could find the dimensions of a figure and then use the formula to find the area. She identified that approach as analytic. She then described an analog approach as reading the information on a bag of sand to see that it covers so many square feet. Another student shared yet a different image of the problem, suggesting that "two Silver Stadiums equals one War Memorial" (two local sports arenas) on the basis of the number of people each could hold.

The group then began to struggle with understanding the conditions for the theorem that was required—a theorem about two different figures that have the same area. Someone offered, "Maybe like a 2×3 square equals a 6×4 triangle?" Another person wondered what an analog approach to finding the area could be, and someone else continued to talk about the idea of separating a figure into two parts and then adding them.

It is important to note that although this small group had no facilitator, the students initially engaged in a process similar to the one used to understand the previous essay. They first spontaneously shared their individual interpretations of the task—much like Tim and Char initially did—which allowed them to develop a shared sense of the task. Then again like the previous experience, they began a group effort to understand the specific details of the task.

Group 2—deciding on a procedure to carry out the task

The group decided to take a 4 × 4 square and make a rectangle with the same area. As one person began to measure and draw the square, the following interchange took place:

Tim: I've got to measure it, four inches by four inches.

Shellie: What can I do?

Eric: You can make the rectangle.

Shellie: Okay, what size?

Tim: We're not sure yet.

Shellie: How you gonna make a square fit in a rectangle? (*jokingly*) Two squares equals one rectangle?

Tim: I'm not sure.

Eric: No, one square equals one rectangle.

Shellie: It can't. A rectangle's big.... Oh, the area.

Eric and *Margie:* Yeah !

(This entire small-group discussion is a good example of *students communicating to students—about technical mathematics and about their process of doing mathematics—through discussion.*)

Tim finally said, "Maybe a 2 × 6 would be adequate, yeah, a 2 × 6 rectangle." When asked how he decided, he said that he was guessing and then went on to explain *how* he was guessing.

Tim: Take two inches off the top and that equals 6, and this is down to 2, so then it would be equal in area.

 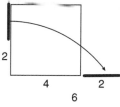

Tim: (*He elaborates and revises his original idea when Shellie says that she does not understand it at all.*) Actually, I think cutting the square in half would be analytic, and analog is just adding these two together; cuz, go take two off here and put it over here, see.... Wait a minute.... If you take this top and put it over here.... Ohhh.... That'll be eight. Should have been

8, a 2 by 8.... (*student communicating to self and student communicating to students—about technical mathematics—through drawing and telling and showing*)

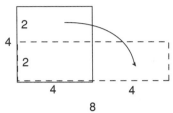

They proceeded to carefully measure, draw, and cut out their figures, while continuing to talk about whether the rectangle was going to fit into the square or vice versa. Ultimately they decided that it did not matter and followed through with their plan.

Tim: We're gonna make this rectangle fit inside this square.

Shellie: It's gonna cuz this little piece can go right next to it. (*To Eric*) Don't cut it!

Eric: Have to.... (*He works on cutting the rectangle.*) Now, hopefully, this little.... (*He places it on top of the square.*)

The group: (*Joyously*) It fits! (*students to students—about technical mathematics—through a demonstration*)

It is especially interesting to notice how these students are interacting. The combination of the negotiations that are taking place in conjunction with the different modes that students are using to communicate (i.e., discussions, telling/showing, drawing, and demonstrating) is allowing them to create their results. Notice also that because these students are explicitly talking about the processes they are using, not only are they able to self-correct but, in the end, Shellie, who intially asked the most questions, did not even need to cut the rectangle to know that it would cover the square.

Group 2—creating the theorem

Now this group was ready to create its theorem. Group members spent a great deal of time trying to determine just what a theorem is and ultimately decided to read a few theorems from a textbook to get a feel for what they are like. They tried to verbalize their work, using the same tone as the textbook. They began with this sentence: The area of a square equals the area of a rectangle. As students continued to discuss the nature of theorems, that they seem to apply in general and that they have to be understood by an outsider, they created several iterations of their own theorem. Each time they identified a weakness in their statement, they tried to improve on it, finally settling on the following:

> If the length of side a on the square multiplied by the length of b at the top of the square equals the length of side c on the rectangle multiplied by the length of side d at the top of the rectangle, then the rectangle and the square have the same area.

It is interesting to note that in the group we have chosen to portray here, no one person knew how to complete the entire task. Yet as they listened to and questioned one another, they were able to construct a process for completing the task and producing an appropriate product. Although this product may, on the surface, look rather unsophisticated, the mathematical concepts they reconstructed—the conservation of area, the nature and role of theorems, and the development of a process to calculate the area of a square and a rectangle—are quite sophisticated.

Event 8: The groups synthesize their work and communicate the results of their inquiry to the whole class

Each group documented and shared its theorem and proofs, explaining its analog and analytic approaches. Group 2, for example, stated that its analytic approach was using the formula to generate some concrete examples of different figures having the same area so that the group could come up with a theorem; the analog approach consisted of cutting out the pieces and showing that the pieces covered the same surface.

Notice how the need to communicate the results of their work to the class gave the students a genuine reason to clean up, organize, and reflect on their work. The nature of this communication is also quite different from that of the communication that took place while the group created its results. There is no more negotiation or generation of ideas; instead, the students are presenting the results. Even so, by sharing the different results from each group, the students were able to see that there were multiple ways to carry out the task and multiple theorems that could be developed.

Event 9: Completing the inquiry by creating a context for their theorem

To help the students more fully appreciate the benefits of studying mathematics from a historical and social perspective, the teacher also asked each student to create a context for the theorem their group had created. In preparation, the students read several snippets to remind them of some of the many different times and places in which the Pythagorean theorem was discovered as well as to act as a model for writing about their own theorem. Students demonstrated a great deal of creativity in their snippets, which once again were shared with the whole class. Since it was now the last day of school before a long vacation, the experience was drawn to a close with only a brief reflection on the entire process of constructing their own mathematics.

Notice that here again, as was the situation with the culmination of the essay experience, the students have shifted their conversation to a different level. Here it has shifted from students' presenting the specific details of their individual theorems to their recognizing and appreciating the fact that they have actually constructed their own mathematics.

WHAT THIS EXAMPLE TELLS US ABOUT COMMUNICATION IN A MATHEMATICS CLASSROOM

The previous narrative illustrated the complex weaving of multiple contents of communication, multiple modes of communication, and multiple configurations of participants that can take place in a mathematics classroom guided by a spirit of inquiry. We now briefly share our responses, from the perspective of the teachers of this course, to the main questions we raised about communication at the beginning of the chapter: In what ways can communication take place in a mathematics classroom? Who is involved in the communication? What can be the content of the communication ?

If communication is thought of as sending messages, then this example has shown that several modes of communication can be used in a mathematics classroom. More specifically, the analysis of the previous narrative led to the identification of the following seven categories with respect to modes of communication: reading; writing; telling/showing; modeling/demonstrating; discussion; drawing; and viewing/observing. These categories are first of all characterized by the use of a different medium of communication (i.e., written text—reading and writing; oral verbal conversation—telling/showing and discussion; images—viewing/observing and drawing; actions—modeling/demonstrating). Even when the same medium is used, these methods of communication may be further distinguished in terms of whether they mainly involve receiving versus generating messages (e.g., reading versus writing; telling versus discussing; viewing/observing versus drawing), although in either situation the process would involve a construction of meaning from all participants in the communication (see Rosenblatt [1994]).

Although the narrative reported in the previous section and table 20.1, which summarizes the occurrences of each of these modes of communication, illustrate how this example involved instances of communication that used all these modes, this is not usually the case in traditional mathematics classrooms, where communication occurs predominately in the form of telling/showing and, to a lesser extent, writing and reading. In contrast, this example shows several instances of genuine discussions, where the whole learning community engaged in the negotiation and generation of ideas. This example also showed the value of modeling/demonstrating where the teacher or a student not only communicates the end product of his or her thinking or activity but also the process that led to such a product. Furthermore, it illustrates how the same mode of communication can take on different forms and roles in an inquiry-oriented mathematics classroom (Lampert 1990; Ball 1991). For example, in this experience, telling and showing often gave the students, not only the teacher, a means to share what they knew with the rest of the community; students read a variety of mathematics-related texts and approached them in a generative way, which allowed them to go beyond simply extracting information (Siegel et al. 1996) as is typically the situation when students read textbooks individually. Writing was often used not simply as a way to communicate already made meaning but also as a tool for thinking (Emig 1977; Rose 1988).

This example also illustrates that many different reasons to engage in each mode of communication may be identified in the context of mathematics instruction and demonstrates that these reasons determine the impact and value of such communications for students' learning. For instance, drawing was used to make sense of a reading, to share one's understanding, and to visualize a physical example; reading was used to validate students' conceptions, to gather information, to share one's thoughts, and to search for a model; writing was used to acknowledge and communicate about one's feelings, to model the sorting of verbal information, to act as a visual prompt, and to communicate the results of the group's work. Clearly, effectively increasing the ways in which we communicate in a mathematics classroom calls for both students and teachers to become more aware of the powerful role the purpose for communicating plays in determining the mode of communication.

Although it is clear from this narrative that the portrayed class functioned as a community of learners and that all communication was intended for the benefit of the entire community, it is possible to identify at least four configurations with respect to who was primarily involved in the communication in each specific instance. We have characterized these configurations as follows: student(s) to self, student(s) to teacher, teacher to student(s), and student(s) to student(s) (a label that we also used to describe conversations involving the whole community, including the teacher). Note that in some occasions, these interactions were not direct, but rather occurred through the use of a written text or video that had been chosen by one of the parties as a means to communicate a specific message.

Analyses of the discourse typical of traditional mathematics classrooms (Bishop 1988; Hicks 1995) have pointed out how teacher-to-student(s) interactions are predominant in this situation. In contrast, a look at table 20.1 shows that in this example, all four configurations occurred consistently throughout the experience, with student-to-student communications being the most predominant and occurring in every single event. More generally, we can say that in this instructional experience, it is the students' voices and ideas that we hear the majority of the time.

Finally, when we look at the content of the communications that took place in this example, five categories emerge: technical mathematics (i.e., dealing with specific mathematics concepts or techniques); the nature of mathematics (i.e., dealing with issues "about" mathematics as a discipline or metamathematical concepts); the process of doing mathematics (i.e., how one solves problems in mathematics); how one is learning (i.e., dealing with the process of learning or particular learning strategies); and feelings about what one is learning. Once again, this list expands considerably what are considered legitimate topics of conversation and learning in a traditional mathematics classroom, which is usually limited to technical mathematics. Despite the many recent documents calling for changes in what school mathematics we teach and how we teach it (see National Research Council [1989]; NCTM [1989, 1991]), it is still very unusual to hear mathematics classes talk about the nature of mathematics, about how they read an essay, or about how they feel about the information being presented. It seems, however, that these are precisely the kinds of things that students can and must talk about if we want them to develop a view of mathematics as a way of thinking and to see themselves as capable of doing mathematics.

FINAL REMARKS

This example has shown that communication can play a powerful role in a mathematics classroom. It has demonstrated that students can, and will, talk about mathematics when given the opportunity and support to do so. Further, it has shown that there are many ways to communicate and there is more than just technical mathematics to communicate about. As teachers, we need to recognize that how we communicate may be as important as what we communicate. We need to begin to orchestrate classroom experiences that go beyond telling about and showing specific mathematics concepts and techniques to invite all the members of the learning community to engage in interactive communications about mathematics and about learning mathematics.

REFERENCES

Ball, D. L. "Implementing the Professional Standards for Teaching Mathematics: What's All This Talk about 'Discourse'?" *Arithmetic Teacher* 39 (1991): 44–10.

Bishop, A. *Mathematical Enculturation.* Dordrecht, Netherlands: Kluwer Academic Publishers, 1988.

Borasi, R. *Learning Mathematics through Inquiry.* Portsmouth, N.H.: Heinemann Educational Books, 1992.

Borasi, R., and M. Siegel. "Reading, Writing, and Mathematics: Rethinking the Basics and Their Relationship." In *Selected Lectures from the 7th International Congress on Mathematical Education,* edited by D. Robitaille, D. Wheeler, and C. Kieran, pp. 35–48. Sainte-Foy, Quebec: Les Presses de l'Université Laval, 1994.

Davis, P., and R. Hersh. *The Mathematical Experience.* Boston: Houghton Mifflin, 1981.

Emig, J. "Writing as a Mode of Learning." *College Composition and Communication* 28 (1977): 122–27.

Harste, J., and K. Short, with C. Burke. *Creating Classrooms for Authors.* Portsmouth, N.H.: Heinemann Educational Books, 1988.

Hicks, D. "Discourse, Learning and Teaching." In *Review of Research in Education,* edited by M. Apple, pp. 49–95. Washington, D.C.: American Educational Research Association, 1995.

Lampert, M. "When the Problem Is Not the Question and the Solution Is Not the Answer: Mathematics Knowing and Teaching." *American Educational Research Journal* 27 (1990): 29–63.

National Council of Teachers of Mathematics. *Curriculum and Evaluation Standards for School Mathematics*. Reston, Va.: National Council of Teachers of Mathematics, 1989.

————. *Professional Standards for Teaching Mathematics*. Reston, Va.: National Council of Teachers of Mathematics, 1991.

National Research Council. *Everybody Counts: A Report to the Nation on the Future of Mathematics Education*. Washington, D.C.: National Academy Press, 1989.

Project Mathematics! *The Theorem of Pythagoras*. Videotape. Pasadena, Calif.: California Institute of Technology, 1988.

Rose, B. "Writing and Mathematics: Theory and Practice." In *Writing to Learn Mathematics and Science*, edited by P. Connolly and T. Vilardi, pp. 15–30. New York: Teachers College Press, 1988.

Rosenblatt, L. "The Transactional Theory of Reading and Writing." In *Theoretical Models and Processes of Reading*, edited by R. Ruddell, M. Ruddell, and H. Singer, pp. 1057–92). Newark, Del.: International Reading Association, 1994.

Siegel, M. "Reading as Signification." Ph.D. diss., Indiana University, 1984. Abstract in *Dissertation Abstracts International* 45 (1984): 2824A.

Siegel, M., R. Borasi, and C. Smith. "A Critical Review of Reading in Mathematics Instruction: The Need for a New Synthesis." In *Cognitive and Social Perspectives for Literacy Research and Instruction*, edited by S. McCormick and J. Zutell, pp. 269–77. Thirty-eighth Yearbook of the National Reading Conference. Chicago: National Reading Conference, 1989.

Siegel, M., and R. Borasi. "Demystifying Mathematics Education through Inquiry." In *Constructing Mathematical Knowledge: Epistemology and Mathematics Education*, edited by P. Ernest, pp. 201–14. London: Falmer Press, 1994.

Siegel, M., R. Borasi, J. Fonzi, L. Sanridge, and C. Smith. "Using Reading to Construct Mathematical Meaning" In *Communication in Mathematics: K–12 and Beyond*, 1996 Yearbook of the National Council of Teachers of Mathematics, edited by Portia C. Elliott, pp. 66–75. Reston, Va.: National Council of Teachers of Mathematics, 1996.

Siegel, M., and J. Fonzi. "The Practice of Reading in an Inquiry-Oriented Mathematics Class." *Reading Research Quarterly* 30 (1995): 632–73.

Epilogue

OVER the past decade, research in mathematics education has increasingly turned to analyzing everyday mathematical teaching and learning processes. When Austin and Howson were writing their review of literature on language in mathematics education in 1979, the focus of most of the papers was on the written language of mathematics—on the specific mathematical symbolism and mathematical texts. Today, the review would revolve around the problems of mathematical discourse rather than mathematical language and on the processes of interactive communication in the mathematics classroom. Instead of looking at the readability of different kinds of texts by analyzing their structure and linguistic features, we are more interested in how a student reads a text and what the relations are between the structural and linguistic features of a text and the student's strategies of reading and understanding the text. Problems of communication prevail over those of language alone. The trend of these changes is well exemplified in the chapter on the evolution of views in the German research in mathematics education. It is also discussed in Kunze's chapter, from the point of view of a Wittgensteinian philosophy of language.

Earlier studies often more or less openly implied that successful communication is a matter of using an appropriate language, a well-chosen metaphor, or an illuminating visual representation. There was a hope that the reality of mathematics classes could be changed at will by means of finely structured and methodologically sophisticated didactic materials. Newer analyses suggest that the complex interactions among the social, psychological, communicational, and mathematico-epistemological aspects in the teaching and learning processes are subject to independent "laws" that relativize the benefits of any single solution to the problem of communication in mathematics.

Research is now at the stage of trying to understand the complex phenomena of language and communication in the domain of mathematics. There is a growing awareness that the view of mathematics as a syntactical structure for denoting and studying relations between objects and magnitudes is very far from what functions as mathematical knowledge in the processes of learning mathematics at school, in the processes of its creation by mathematicians, and in its use by those who apply

341

it in their professions. From the latter point of view, mathematics is more like a living organism, a language-in-use, a culture. Thus, learning mathematics becomes an initiation into a very special culture in which quite a few things are taken for granted and very many things cannot be taught explicitly and directly.

The observation of communication in mathematics classes has led to the identification of certain patterns of interaction between teachers and students that may well teach the students how to survive in these classes but do not help them become active participants of the mathematical culture enjoyed by its creators and users. Routines and stereotyped frames of communication, like the well-known initiation-response-feedback or funnel patterns, leave the students speechless mathematically, although they keep the appearances of an exchange of ideas. The students learn the implicit conventions of such apparent mathematical communication, but these conventions are irrelevant from the mathematico-epistemological point of view. However, this state of affairs may not be easy to change. The standardization of interactions and the superficiality of the communication may well lie in the very nature of institutionalized learning and teaching. The instances of mathematically rich and meaningful communication reported in this book all show at how high a cost they have been obtained, in terms of the teacher's involvement, the students' intellectual and emotional investment, time, and organizational skills.

There is certainly a better understanding and awareness now than in 1979 of what is going on in the exchanges between teachers and students and among students themselves as they work in groups. But there is also a growing need for theories, analytical tools, concepts, and research methodologies that would allow one to capture the subtleties and complexities of the related phenomena and to be able to speak about them. The variety of theoretical frameworks and methodologies represented in the preceding chapters testifies both to this strife and to the lack of a more general methodology that would apply to the study of a wider range of situations.

The perspectives assumed by Cestari and Bartolini Bussi look at the processes of teaching and learning from two complementary vantage points: social-constructivism and activity theory. Kanes introduces a Wittgensteinian approach in which the language itself constitutes instruction, and Seeger takes a broader view in which the introduction into a culture involves simultaneous participation in its mimetic, oral and narrative, and theoretic levels. Each of these levels has different cognitive consequences. Seeger, in fact, raises the question of whether the focus on the mathematical discourse in research is not giving too partial an account of the processes of teaching and learning mathematics.

Wood and Krummheuer take an interactionist perspective and discuss different interaction patterns and formats. But even here, there is a difference. The two authors seem to have different ideals in mind. Whereas Wood stresses the need to allow a greater freedom of expression in students and views the teacher's role mainly as focusing and sustaining students' attention on certain aspects of a mathematical situation, Krummheuer sees the interaction in a teaching-learning situation as much more constrained by the mathematical contents. The teacher formats the interaction in a certain way to achieve certain well-defined objectives.

The chapters in our book contain long excerpts from transcripts of verbal interactions, but it is clear that tools for their analysis are in great need of development or improvement. Mathematics education research is slowly learning from discourse analyses, as illustrated in Cestari's chapter, but as Bartolini Bussi points out, it is increasingly obvious that original methods, adapted for the purposes and interests of this particular domain, must be devised. Discourse analysis in mathematics education must take into account the specific mathematical content of the conversations, and its basic unit of analysis cannot be a word or even an utterance but an episode whose unity is decided by one common mathematical or metamathematical question or problem. Therefore, not looking to discourse analysis as a sole source of inspiration and model, mathematics educators use a wide variety of approaches, analytic tools, and ideas. This variety is clearly in evidence in our book.

Cestari's analysis of protocols is overtly inspired by conversational analysis and discourse analysis. Bartolini Bussi proposes a methodology of transcript analysis influenced by Vygotskian ideas. For the purposes of research in mathematics education she finds it necessary to distinguish between coarse-grain and fine-grain as well as long-term and short-term analyses of discourse. Kanes demonstrates a methodology of protocol analysis based on Wittgenstein's conception of language games as a tool for analyzing language. But Kanes also raises a question: To what purpose should such analyses be put in mathematics education? Do we just want to understand how mathematical meanings emerge within classroom interactions and what kind of mathematical meanings they are, or do we aim at bringing about changes in these processes?

Chapters on small-group interaction in part 3 are particularly concerned with the problem of finding adequate analytical and methodological tools for the study of mathematical communication. These tools must allow one to answer such specific questions as these: How can we identify those features of a small-group interaction that make it fruitful in the sense of the growth of understanding in the participants? What forms of teacher-student interaction are more likely to enhance students' understanding of mathematics? The chapters by Curcio and Artzt and by Stacey and Gooding propose two instruments for evaluating mathematical discussions in small groups. One is more qualitative, inspired by Schoenfeld's work on metacognition in problem-solving contexts; the other combines the quantitative and qualitative approaches on two coding schematas of the verbal interaction in small groups.

The book leaves us with many issues that need to be addressed. The problems of more comprehensive frameworks and more adequate methodologies and the question of the mechanisms of effective communication in small-group discussions are but a few.

A very important problem is brought forth by Cestari's analysis of a few episodes of teachers' interactions with students in the frame of a Brazilian project, "Aprender pensando," deemed to create a constructivist learning environment. This analysis shows how difficult it is for a teacher to change his or her patterns of communication with the students. The ideology of constructivism, underlying the project of reform, created an idealistic vision that did not take into account the actual processes of mathematical meaning-making in the everyday life of classrooms.

A similar point is, in fact, made in Civil's chapter, which brings to our awareness the phenomenal difficulties in the mathematical training of elementary school teachers. These prospective teachers have already as students been enculturated into some forms of mathematical communication. These forms of communication have become inseparable from their mathematical knowledge. Is it possible to format the interactions with them and among them in a way that requires overcoming these older habits and beliefs? Civil raises yet another quite important issue in relation to the constructivist style of teaching. Overcoming sociocognitive conflicts is deemed to enhance learning and development. But is it fair to push students working in a group into addressing a conflict situation brought about by a peer's (erroneous) mathematical strategy? Can students be left with the responsibility of helping a peer in overcoming a conceptual difficulty whose reasons are often unclear even to researchers in mathematics education?

A few examples of successful communication are found in the book together with an analysis of what accounts for the success. But does it make sense to speak about "successful communication" in mathematics from the point of view of constructivism, sociocultural approaches, or interactionism? Are we fated with relativism or are there values on which a consensus could be obtained? Given a certain ideal model of mathematical communication, what are the conditions of its realization in practice? Can these conditions be satisfied in view of the "laws" that govern institutionalized teaching and learning? Are we sure that we know what we mean by "clarity" in a mathematical explanation? Kanes asks, What is the role that "clarity" plays in mathematics teaching?

Concerning successful communication, Steinbring in his chapter reminds us of the warning that comes clearly from Voigt's research on patterns and routines of communication: successful communication is no indication of successful learning of mathematics. Students may be successfuly learning only the ritual of interaction with their teacher, which is why in-depth epistemological analyses of the mathematical content of classroom interactions are needed. Remarkably, it is also this aspect of research in mathematics education that constitutes its originality and professionalism. These analyses cannot be made by general sociologists, linguists, communication theorists, or discourse analysts.

Quite a few chapters in the book (especially in part 4) look at the specific features of the language of mathematics. It is a very particular language that uses, simultaneously, a wide range of registers and representational systems. In "speaking mathematically," one has to move constantly between these different registers (ordinary language, formal notation) and forms of representation (formulas, graphs). This flexibility is the mark of mathematical thinking and is very difficult to acquire. The learning of mathematical language remains a complex issue, in spite of the growing body of publications on the subject. Algebra, especially, poses problems, and although some time ago a belief arose that picturesque metaphors and the concretization of abstract relations could facilitate students' understanding, today we suspect that as Kath Hart once said, "by trying to 'make concrete' certain parts of mathematics, we have confused rather than helped children." We also have a better understanding of the gap between arithmetical and algebraic thinking and how children and adults are sometimes able to fill this gap by passing through a phase of using a syncopate style of algebraic interpretation of story problems (Arzarello). The question is how they are doing it and in what conditions.

Another major issue is that of the contexts in which mathematics is learned. How does this dimension affect the performance of different groups of students (of different genders or coming from different cultures)? The view that teaching should keep to contexts that are familiar to the students, quite popular a while ago, is challenged in the chapter by Clark. Although there is no doubt that students do better on problems worded in ways that evoke familiar contexts, it may be wise to convince students to widen the scope of contexts with which they are familiar. It is also important for educators to keep up with the new trends in the interests of different groups of students. Knitting, traditionally regarded as a favorite female pastime or occupation, may today be familiar to a very small percentage of teenage girls. Keeping up with modern times, in general, should be a principle in education, but it is not always respected. Pirie brings to our attention the fact that the balance metaphor, used to teach solving linear equations, is conveyed in textbooks through drawings of old-fashioned scales that in many places of the world can today be seen only in a museum.

Many different views are represented in the book, but all authors share the belief that mathematics is a human construction, a part of the human culture. The meaning of mathematical symbols and terms is nowhere reduced merely to their connotations, provided by definitions and circumscribed by theories, but is related to situations in which these symbols and terms are seen as playing a structuring role. How meaningful objects (also mathematical symbols and terms) structure the given situation for the learners and the teachers depends very much on the standards or conventions of the interpretation of signs in this type of situation (for example, a mathematics lesson), and these are not merely linguistic: they are cultural.

This book has brought together authors from very different backgrounds, who look at the problem of language and communication in the mathematics classroom from a wide variety of points of view. This by itself is an interesting feature of the book and furnishes material for further study.

As we communicated in our meetings in Quebec, the differences among our mother tongues and accents in English obviously created difficulties. Is it surprising that many of these difficulties were overcome by our "taken as shared" understanding of the mathematical language, which was a mother tongue to no one?

REFERENCE

Austin, J. L., and A. G. Howson. "Language and Mathemetical Education." *Educational Studies in Mathematics* 10 (1979): 161–97.

Index